Standards for Open Systems Interconnection

STANDARDS FOR OPEN SYSTEMS INTERCONNECTION

K. G. Knightson
T. Knowles
J. Larmouth

McGraw-Hill Book Company
New York   St. Louis   San Francisco   Colorado Springs
Montreal   Oklahoma City   San Juan   Toronto

Library of Congress Cataloging-in-Publication Data

Knightson, K. G.
    Standards for open systems interconnection.

    Includes index.
    1. Data transmission systems — Standards.
I. Knowles, T. II. Larmouth, J. (John) III. Title.
TK5105.K59 1987   004.6'2   87-26068
ISBN 0-07-035119-8

1234567890   DOC/DOC   89210987

ISBN 0-07-035119-8

This book was first published in Great Britain in 1987 by BSP Professional
Books, Osney Mead, Oxford.

Printed and bound in the United States of America by R.R. Donnelley & Sons
Company.

Contents

Foreword   viii
Dedication and Acknowledgements   ix

Part 1 Overview and Background
  1 Philosophy and Background   3
    1.1 Background   3
    1.2 Aims   7
    1.3 Practice   9

  2 Standards Organisations and the Standards Processes   12
    2.1 Introduction   12
    2.2 The International Standards Organisation   14
    2.3 International Telephone and Telegraph Consultative
         Committee   16
    2.4 European Computer Manufacturers' Association   17
    2.5 Liaison and co-operation   18

  3 The Reference Model   20
    3.1 Layer principles   20
    3.2 The concept of a Layer   26
    3.3 Layer service principles   29
    3.4 General addressing principles   36
    3.5 Relations between Layer connections   39
    3.6 Other common elements of Layer operation   41
    3.7 Relaying   48
    3.8 The connectionless Addendum   50
    3.9 The naming and addressing Addendum   54

Part 2 Interconnection of Open Systems

  4 Layers 1 to 3   59
    4.1 General requirements of the Network Layer service   59
    4.2 Subnetwork interconnection   60
    4.3 Internal Organisation of the Network Layer (IONL)   61
    4.4 OSI Network service and its provision   68
    4.5 The OSI connectionless-mode network service (CLNS) and its
         provision   105
    4.6 Network interconnection   117
    4.7 Network Layer addressing   121

4.8	Data link Layer	132
4.9	Physical Layer	145
5	**Transport Layer**	**154**
5.1	Transport service	156
5.2	Transport protocol	164
5.3	Addenda and enhancements	176

Part 3 Communication Between Open Systems: Common Services

6	**The Session Layer**	**181**
6.1	Introduction	181
6.2	The Session service	183
6.3	The Session protocol	200
7	**Overview of Presentation and Application Layers**	**206**
8	**The Presentation Layer**	**210**
8.1	Introduction	210
8.2	Presentation Layer concepts	213
8.3	Abstract Syntax Notation	221
8.4	Basic Encoding Rules	227
8.5	Summary and conclusions	229
9	**Common Application Service Elements**	**230**
9.1	Introduction	230
9.2	Association Control Service Elements (ACSE)	231
9.3	Commitment, Concurrency and Recovery (CCR)	236

Part 4 Communication Between Open Systems: Specific Services

10	**File Transfer, Access and Management**	**259**
10.1	History and objectives	259
10.2	Philosophy of FTAM standardisation	260
10.3	Overview of FTAM	261
10.4	File contents	265
10.5	The virtual filestore	271
10.6	FTAM service primitives	276
10.7	Constraint sets and document types	284
10.8	The reliable service	288
10.9	Transfer of FTAM documents	291
10.10	Typical FTAM implementations	299
10.11	Conclusion	301
11	**Job Transfer and Manipulation**	**302**
11.1	History and objectives	302
11.2	Use of Commitment, Concurrency and Recovery	303

11.3	Job Transfer and Manipulation overview	303
11.4	Job Transfer and Manipulation subsets	305
11.5	Job Transfer and Manipulation model and terminology	307
11.6	Job Transfer and Manipulation work specification	308
11.7	Error-handling	319
11.8	Authorisation	320
11.9	Job Transfer and Manipulation Service primitives	321
11.10	Job Transfer and Manipulation protocol	324
11.11	Document types	325
11.12	Conclusion	326
12	**Virtual Terminal Service and Protocol**	327
12.1	Introduction	327
12.2	The Virtual Terminal Model	329
12.3	The Virtual Terminal Service	336
12.4	Extended facility set	345
12.5	Protocol	349

Part 5 Wider Issues

13	**Management**	355
13.1	Background and history	355
13.2	The OSI management framework	356
13.3	Common Management Information Service (CMIS)	359
13.4	Fault management	360
13.5	Accounting management	360
13.6	Configuration management	361
13.7	Performance management	361
13.8	Security management	361
13.9	Name registration and management	362
13.10	Conclusion	362
14	**Security**	363
14.1	Introduction	363
14.2	Types of Protection	363
14.3	Protection Mechanisms	365
14.4	Architecture – placement of security services	367
14.5	Non-OSI aspects	368
14.6	System security	369
14.7	Protocol enhancements	370
15	**Conformance, Conformance Testing and Procurement**	372
15.1	Conformance requirements	372
15.2	Conformance testing	374
15.3	Procurement	377
Index		381

Foreword

Open Systems Interconnection (OSI) is a significant step towards the realisation of the rewards anticipated from Information Technology.

The development of OSI has involved thousands of experts from all over the world in efforts sustained over a period of seven years or more.

We are just beginning to be able to taste the fruits of these efforts as implementations of OSI protocols are being put into daily use. Large organisations such as General Motors, USA, are sponsoring and promoting OSI for their own purposes and manufacturers are producing products conforming to OSI.

This book is intended for anyone who wants answers to questions such as:

- What is OSI?
- Why might I wish to use OSI?
- How do I go about implementing OSI?
- How should I approach procurement of OSI implementations?

Unfortunately, the subject matter of this book is a moving target of significant standards work in progress.

Whilst the authors have been at pains to ensure that it is accurate at the time of writing, the time taken to publish the book will inevitably see some changes to the standards described.

Nevertheless, most of the work described is in its final stages and the nature of such changes are likely to be minor. The authors feel therefore that the time is right to provide explanations of and insight into the OSI communications standards because products are under development and user organisations will be planning future strategies based on OSI.

Dedication and Acknowledgements

The authors wish to acknowledge the often considerable efforts of all those involved with the development of OSI, without whom this book could not have been written.

The authors also wish to thank all those who have helped in the writing and publishing of this book.

Particular thanks go from Terry to Joan K. Gibson, his wife, whose dedication, considerable assistance and encouragement saw this book through from its early days to its hectic final stages, and from John to his wife and the twins Sarah Jayne and James Henry.

PART 1

OVERVIEW AND BACKGROUND

CHAPTER 1
Philosophy and Background

1.1 BACKGROUND

The past twenty years have produced a revolution in the computer and communications fields and in the past ten years these two fields have become interlinked.

The technology of the computer has become the basis for new Stored Program Control (SPC) switching equipment used for communications and the increasing use of computers has produced a demand for data communications.

This data communications demand is witnessed by the many national public data networks that have been deployed since the early 1970s. Most of these data networks offer CCITT X.25 (or a variant of it) as the interface for computers and X.28 as the interface for simple asynchronous terminals. These are examples of data communications standards of which more will be said in Chapter 4.

Data communications standards are needed essentially because of the need for communication between equipment from different manufacturers. A standard is basically a design or 'blue-print' for construction of a piece of equipment (or part of) which is agreed between a number of interested parties. Communications standards are of interest to manufacturers, users of computer equipment, PTTs and software houses. All of these have participated in the production of Open Systems Interconnection (OSI) standards.

The earliest data communications standards arose from telegraph and telex applications. These standards defined signals and character sets in order that messages could be exchanged between humans. The character sets were orientated around the 5-hole paper tape used on 'ticker-tape' machines. Transmission errors were detected mostly by the humans reading the received messages.

As the more sophisticated 'teletype' was introduced for telegraphy, a 7-bit character set known as ASCII was defined by the USA standards organisation, ANSI, and later adopted by ISO as the basis for ISO 646. This character set includes the basic alphanumeric and punctuation characters used in the English language. It also contains some non-printing control characters with meanings related to the definition of the start and end of messages and message headers (STX, ETX, SOH, etc.), synchronising characters for synchronous transmission and flow control characters allowing receivers to halt and resume the flow of data.

The characters are normally transmitted and stored as 8-bit characters, the eighth bit being a parity bit so that simple 1-bit errors can be detected automatically. Some problems were found with the ASCII character set which serve to illustrate some of the areas of concern of the OSI standards.

Firstly, the character set defined by ANSI was orientated around the needs of the USA. Thus it contains a '$' but no other currency symbol. Also it does not contain accents or other symbols needed by non-English languages and scientific texts. As a result there are many national variants of this character set. Even between the UK and USA there are differences. It is common for '£', '$' and '#' (hash) to be given different codes and for any single piece of equipment to be able to print or display only two out of these three characters.

Although ISO has now standardised each national character set, the problem still remains in practice because each piece of equipment often supports only one national character set.

Secondly, the standard allows message content and message control functions to be mixed in such a way that transparency is not possible. 'Transparency' here means that user data to be transmitted must not contain any bit pattern corresponding to any member of the set of control characters. As a result the transmission of some types of data is either very inefficient or impossible. For example, consider the number 65535. To transmit this value, using the ASCII character set, requires five characters, one for each decimal digit. If any binary bit pattern could be transmitted, this number could be transmitted in just two 8-bit characters.

The lack of true standardisation of the ASCII character set and the mixing of essentially different types of function (information representation and transmission control) have over the years become recognised as serious problems.

There are other problems with the early methods used for data communication and these also stem largely from the limited objectives set by the protocols designers, limitations imposed mostly by limitations in available technology and by restricted user requirements.

To understand these limitations it is necessary to examine the state of the industry ten to twenty years ago.

At that time, computers were large and expensive and employed either for exotic research or by very large organisations for administration. They were exclusively a centralised resource, initially used for batch processing. Work to be performed was punched out on to cards or tape and submitted to the computer centre where it joined a queue of work. Eventually the job reached the head of the queue, was processed and the results printed.

This method of working was designed around the needs of the computer and was intended to optimise the use of the precious computer resource. The problems for the users were manifold. Jobs frequently took twenty-four hours to be processed and all too often the end-result was a list of errors in the input which prevented the job from being performed to completion.

As computers became less expensive, use of computers to validate data as it was being typed became possible and in some cases on-line access to interactive enquiry facilities was possible.

It was at this time that data communications started to grow. Computers were still relatively costly compared to other resources and still centralised. There was a need to distribute the human interface equipment (terminals and remote job entry stations) to the various places where they were needed. Hence, networks were needed.

Communications networks were exclusively star networks, all links going from terminals to the central computer. Terminals were very 'dumb' and the central computer contained all the 'intelligence'. The protocols used were designed around these factors. They were asymmetric or 'master–slave' protocols where terminals were told by the central computer when data could be sent and received. In many cases the terminal status 'locks' were also centrally controlled. The central computer was able to lock the keyboard and prevent any operator action when required. Examples of these types of protocol are the early versions of the IBM 3270 terminal protocol and the ICL 7181 terminal protocol.

At the time, these protocols served their purpose well but technology has once again moved on and changed the requirements.

The microprocessor and the use of VLSI generally has made computer power very cheap. Micros are being employed extensively in manufacturing processes and commerce. They are also to be found in a growing number of homes as games computers and in some domestic appliances. Parts of this book were typed on to a microprocessor on trains and aeroplanes.

The large centralised computer still exists but it does not have the monopoly on intelligence. Some terminals today have more processing power than the mainframe of twenty years ago.

Processing power is being positioned where it is needed. The communications requirement is changed. Instead of simple terminals needing to provide a human interface to a centralised processing resource, distributed processing resources need to co-ordinate their actions.

Also the new technology is making new communications networks available. The bandwidth achievable on land lines is increasing to the point where up to 2 Mbps is now on offer to subscribers. Satellites, microwaves, cellular radio and even lasers are offering alternatives to the land line for wide area communications. The local area network (LAN) offers promise of very cheap high-bandwidth communications within an office or factory complex. These changes all lead to emergence of new styles of systems where intelligence is widely distributed and centralised resources are just those offering specialised facilities such as bulk storage or very high-speed bulk printing.

There will not be a clear 'master' or 'slave'. There will instead be a number of multi-function work stations and a number of 'servers' offering a variety of services on demand to users. In some cases a server will also be a user of services from another server. An example of this could be a computer on a LAN offering electronic mail services needing to access another computer on the LAN for directory services in order to find the physical station address of the addressee of a message.

This all adds up to the fact that the master–slave types of protocol are no longer adequate to the needs of today's computer users.

One further trend is worth noting. The data communications scene of ten years ago was dominated by the manufacturers of the large computers which served at the centre of the star networks. Each manufacturer defined proprietary protocols to suit his own computers and also manufactured the terminals. In most cases the protocols were unique to the type of device being attached so that a paper tape reader used a different protocol to a card reader, barcode reader or terminal, etc.

IBM was the first manufacturer to realise that the incompatibilities within its own product range was a major support problem and would prevent the implementation of the types of distributed system needed for the future.

IBM's answer was the introduction of a communications architecture, Systems Network Architecture (SNA), which attempted to separate logically distinct functions and provide a framework within which these functions could be assembled into an integrated whole.

At the time of first introduction, SNA was still orientated around the master–slave type of operation but the newest version available in 1984 now introduces 'peer-to-peer' protocols where each communication system is regarded as having equal status as far as the protocol functions are concerned. This does not mean that they will be equal in processing power or other functionality, only that neither exerts overriding control over the progression of an instance of communication.

Other manufacturers have also introduced communications architectures so that, whilst we now have a greater degree of rationalisation within the product ranges of a single supplier, the products of different suppliers are usually incompatible.

These incompatibilities are often welcomed by suppliers who are pleased that when a customer has purchased equipment from them there is no other supplier who can provide compatible equipment. The user is 'locked-in' and must keep buying from the original supplier.

It is against this background that the International Standards Organisation (ISO) began its work in 1979 on what was then known as Open Systems Architecture. This has since been renamed and is now known as OSI.

1.2 AIMS

The fundamental aim of OSI is very simple – to define standards for data communication which will enable computers from different manufacturers to communicate with each other.

However, the preceding section has highlighted some of the problems that have to be faced. OSI firmly faces the future and tries to avoid the limitations of the protocols and standards that have preceded it.

OSI offers exclusively peer-to-peer protocols. It does not support master–slave communications apart from CCR (see Chapter 9) and as yet does not cater for the multi-drop style of connection where many devices are connected to the same land-line.

OSI aims to subdivide the communications functions into logically separate modules and more will be said on this subject in following chapters.

From the preceding subsection it will be clear that there is a need to separate user data from communications control functions. OSI modularity goes much further than this, however.

A distinction has been made between connection between two systems and communication between them. To illustrate this, consider an international telephone call from England to Japan. I can make a *connection* quite easily to Japan if I know a telephone number. I cannot *communicate* over this connection if the person who answers speaks only Japanese because I do not speak Japanese. Thus useful and meaningful communication can only take place if there is a commonly understood language defined in terms of syntax and grammar.

Communication between two people speaking the same language is still impossible if both talk at the same time. There is in natural language what can be termed 'dialogue control' whereby a person who is speaking can indicate via voice inflection and pauses that speaking is finished and the other person may talk.

These examples of connection, communication and dialogue control are taken from direct human interactions and illustrate that there are many independent aspects of communication, all of which must be used in conjunction to communicate successfully.

In this book, we cover interconnection, in terms of the types of networks and relevant standards for services and protocols, in Part 2. The common services for communication, including dialogue control, are dealt with in Part 3. Protocols specific to an application or style of communication are covered in Part 4.

This 'bottom-up' approach is logical from the point of view of describing a layered structure where each layer builds on and adds to the functions of the layer below. However, the reader may choose to move from Part 1 to Part 4 if a 'top-down' approach is preferred!

A further aim of OSI is that it should become available as a usable set of standards within a reasonable time from commencement of the international work.

The process of standardisation is a lengthy one. It took about ten years for a single, relatively simple communications protocol, HDLC, to become a full international standard. Because of this, combined with the desire to modularise into a number of inter-related but independent standards, it was necessary to ensure that the standards could be developed in parallel. As any programmer knows, this is quite common when developing large computer programs. What is needed is a definition of the interfaces between the various modules. In the case of OSI, however, there was a difficulty caused by the next aim stated below.

In order to ensure that the OSI standards are acceptable to equipment manufacturers it was necessary to avoid specifying how the protocols should be implemented. In particular, it would have been unacceptable to specify in detail any internal interface in any way that defines how a system should be constructed internally.

The reason for this a simple one. Each computer range has potentially unique hardware and offers different instruction sets and often different internal representation of values. Examples of the latter can be found in character storage where IBM and others use the EBCDIC character encoding for internal storage whereas ICL and most microprocessors use the ISO 646 character coding (based on the ASCII character set).

Thus efficient implementation of a particular 'interface' to a communications protocol will be quite different on each machine.

All this has led ISO to the development of a number of 'abstract' descriptions which specify essential requirements but make no mention of how these requirements should be met by any implementation.

These are the 'architecture' and 'service' descriptions and they will be described in detail later.

Loosely, the architecture defines a number of specific elements of communication (e.g. data encoding, flow control) and a modular grouping of these elements in such a way that they provide a complete communications service to an application.

Each 'service' definition describes an interface between two such groupings of elements.

The need to provide abstract definitions necessarily means that the style of these documents is rather academic. Mention of practical and real-life objects and operations is studiously avoided, for to make such mention would necessarily allude to specific implementations.

OSI architecture and service documents therefore start by defining words like 'entity' and 'service primitive' and go on to describe properties of and interactions between them.

Many readers want to know what these concepts relate to specifically. They are asking the wrong questions, however. To really make use of these standards, one should start by trying to decide what one would *like* them to be in terms of a given machine or system and then check that the specifications fit this desired usage.

1.3 PRACTICE

OSI has been, and still is, a very ambitious project. It requires that a complex set of definitions of interacting pieces of equipment be designed by committee. The reader may be familiar with the old adage that a camel is a horse designed by committee.

In the case of OSI, the committee is, in fact, several international committees which meet at six- or nine-monthly intervals in whichever country can be persuaded to host them.

Most of us who have participated in this work agree that the end-results could be improved. However, no two members can agree on exactly which way to make this improvement!

This illustrates one of the most important facts about the making of any standard. Every effort is made to reach the very best technical agreement possible. In some cases the efforts have been considerable. At the Berlin meeting in November 1980, the Transport Layer group met from 08.00 to 20.00 every day for a week and many delegates spent long hours in the evening preparing for the debates the following day.

In the end, however, there can be times when technical agreement cannot be reached and in these cases political compromise is the order of the day.

Often this happens when a number of participants have entrenched positions, perhaps as a result of having already made a substantial financial investment and the proposed direction for standardisation would be too incompatible.

At the end of the day, however, this is a reflection of the commercial realities of today. A technically perfect standard which no one will implement is of no use to anybody. A useful standard is one which all the interested parties will implement and use, even if the same function could have been achieved more elegantly.

In fact, it has often been said that a standard must be equally unfair to all parties! This saying has more than a grain of truth in it. If one supplier's designs are taken unchanged as a standard then that supplier has a great and unfair commercial advantage, since he already has products available operating to the standard. The other suppliers have all invested in different products and will have to spend additional money to produce the products to the standard.

One more aspect of standards is worthy of note.

The number of interested parties is very large and the span of potential uses for a standard is often enormous. Thus standards are often very comprehensive and all-embracing.

Thus, it is very often possible that in a specific situation more efficient ways of achieving the desired result can be found. However, the variety of specialised solutions which would result leads to precisely the kinds of incompatibilities which we have experienced and which lead to the desire to develop OSI as described earlier.

Similar situations can be found in other areas of standardisation. Many years ago there was a profusion of electrical plugs and sockets in use. As a boy I grew up in a house which was wired up with at least three different types of power socket: 15-amp, 5-amp and two-pin. As a result electrical appliances were never sold with fitted plugs since whatever plug was fitted by a supplier had a 66% chance of not fitting in a socket in the home to which it was taken.

Now we have a single type of plug and socket – 13-amp with three square pins. If I want to run a table-lamp or a double insulated electric drill I could design a simpler and less solidly engineered plug but I would much prefer to use the standard plug in these situations even though it is 'overkill' for many of the uses to which I put it.

The point is simply that the use of standards may carry an overhead in some circumstances but brings great benefits in the form of greater flexibility and convenience. The existence of a limited number of standard methods of communicating will reduce the number of software products that a supplier needs to support and thus ultimately benefit suppliers and users and should reduce costs. This latter is also a result of opening-up free competition in situations where in the past users were 'locked-in' to a single supplier as described earlier in this chapter.

One final point to be made at this time is that where standards contain many options which were the result of a 'political' compromise, market forces will eventually prevail and there is likely to be a reduced set of common options offered to users by suppliers after the standard has been around for a time.

The Transport protocol offers a good example of this. As the chapter on the Transport Layer will show, there are five classes of Transport protocol, each intended to be used over a different-quality network and each offering unique advantages. This number of classes arose because the PTTs envisaged users connecting to X.25 (packet-switching) or X.21 (circuit-switching) data networks and expected the type of usage of networks to be chiefly of the sort where one connection would serve one application. In particular, they wanted compatibility with their Teletex protocols. The large computer manufacturers, however, saw communications almost exclusively in terms of clusters of terminals connected to host computers and thus the usage to be typically of the kind where several terminals are operated simultaneously over a single connection.

In fact, the trend today is towards workstations attached to local area networks (LANs) and sometimes for the LAN to be attached to a wide area network (WAN), each with a very flexible and dynamic pattern of attachment to other devices such as file servers, directory services or other workstations.

In practice, most manufacturers are offering only three of these five Transport protocol classes. Thus variety is reduced and the interests of both users and suppliers will be served, but in the long, rather than the short, term.

CHAPTER 2

Standards Organisations and the Standards Processes

2.1 INTRODUCTION

In this chapter the major international standards-making organisations concerned with OSI standards, their composition and their standards-making processes, will be covered.

There are a large number of organisations who produce and publish standards. Each has different objectives and consequently their procedures are different from each other. However, in these days of multinational companies and international communications, the ultimate goal of the majority of open-systems suppliers and users is to have standards which are fully agreed internationally across a community of users and suppliers and which are stable in content.

The only standards body with these objectives, and which has procedures and a composition which align with these objectives, is the International Standards Organisation (ISO).

The process of achieving such wide consensus takes a long time (about 5 or 6 years is typical) and the need can often arise for standards to be available in a shorter timescale. Thus, particular communities of interest often set about producing standards quickly. In many cases these can then form the basis for the subsequent international standard. Such standards are generally recognised from the outset as being of an 'interim' nature, pending the production of the eventual international standard.

It must be stated that ratification by a standards body is not all that is required for a standard to be useful. It must also be widely available and used. This may come about in one of three ways. Firstly, suppliers could adopt the standards, implement them and make them generally available. Secondly, it can happen that suppliers do not immediately and voluntarily implement the standards, perhaps because they have already invested in alternative solutions. In this case, implementation will only occur as a result of user pressure reflected in requirements stated to suppliers at time of intended purchase. Lastly, it may happen that although users may desire to use the standards they may not make those standards mandatory purchasing requirements at any time because it is apparent that there are no implementations available from manufacturers. Users often do not want to take on the risk of waiting for the implementation or of being the first customer for it. Manufacturers, on the other hand, will often wait for evidence of user demand before embarking on implementation and do not find this evidence in the user's purchasing requirements. In this case, the deadlock can be broken by initiative from government or a large end-user organisation sponsoring a pilot scheme.

These processes can take a number of years, particularly the implementation stage, and this is especially true for the larger organisations with an extensive product range. However, the benefits of standards, and OSI standards in particular, come from the ability of equipment from different suppliers to have the potential to interlink with ease and this can only happen when equipment is supplied with the necessary hardware and the availability of 'off-the-shelf' software to operate according to the standards.

No discussion of standards production would be complete without mentioning IBM. The predominance of IBM in the Information Technology (IT) industry, with more than 50% of the world market share, inevitably means that IBM technologies become *de facto* standards. Most other computer and peripheral manufacturers have to be able to interlink with IBM equipment to some extent and many choose to be compatible to the extent that software written for IBM machines can be run on their equipment.

Why, then, should there be any non-IBM standards? Why not just adopt IBM protocols as they become available?

The answers to these questions are by no means easy and opinions are sharply divided throughout the industry. Some users and suppliers are of the opinion that there is no need for non-IBM standards. They argue that manufacturers will have to offer IBM compatibility anyway and the production of international standards only makes two sets of protocols which have to be implemented, thus increasing suppliers' costs and ultimately costs to the customers.

Others argue that total reliance on IBM to produce all protocol standards is contrary to the interests of free competition; that once all efforts outside IBM have ceased, then IBM will be free to control the release of its documentation only at a time when to do so would secure the maximum commercial advantage against other suppliers. Such a monopolistic situation would eventually push-up prices to the user.

For the foreseeable future, then, suppliers must tread a careful and tricky line between support of international standards and emulation of IBM.

The following subsections will review the major international standards bodies.

2.2　THE INTERNATIONAL STANDARDS ORGANISATION

ISO is an international organisation established to promote standardisation and related activities worldwide. ISO organisation, membership and procedures are designed to ensure full (or at least substantial) international agreement and consensus on its standards which bear the name 'International Standards'.

There is no compulsion to adopt standards that have been agreed. However, the contribution of the resources needed to make standards is entirely voluntary and the existence of the will to put forward these resources is taken as sufficient indication that the standards are needed and will be used.

Membership of ISO is open only to national standards organisations which themselves do not restrict their own membership to narrow groups within their countries. Other organisations may become liaison organisations, attend meetings, make written contributions, but not vote.

ISO divides the topics of standardisation into a number of areas, each assigned to a 'technical committee' (TC). The technical committee which has responsibility for OSI is TC 97, with a title of 'Information Processing Systems'. TC 97 is responsible for all standards in the area of information processing and this covers computer languages, databases, graphics and communications.

Normally, the scope of a TC is still too large to progress individual standards at meetings, so TCs are further subdivided into subcommittees (SCs).

There are two subcommittees responsible for OSI standards, SC 6 and SC 21. In addition, a further subcommittee, SC 18, is responsible for text and office systems and, in particular, for message-handling protocols. Another, SC 20 is responsible for encipherment and its application, including incorporation of encipherment into OSI protocols.

SC 6 has the title 'Telecommunications and Information Exchange Between Systems'. It has the responsibilities for the lower four layers of the Reference Model, i.e. the Physical, Data Link, Network and Transport Layers. These are the layers concerned exclusively with the transfer of transparent user data between open systems.

SC 21 has the title 'Information Retrieval, Transfer and Management for OSI' and has the responsibilities for the Reference Model as a whole and the upper three layers, i.e. the Session, Presentation and Application Layers. These layers are concerned with the structuring of data exchanges between open systems so as to support distributed applications.

Sub-committees are often still too large and formal for detailed work to be carried out by them and are thus divided further into working groups (WGs). SC 6 has four working groups (one for each Layer for which it is responsible) and SC 21 has six . The SC 21 working groups are:

(a)   WG 1: OSI architecture, conceptual schema and formal descriptions;

(b)   WG 2: graphics (not part of OSI);

(c)   WG 3: database (not part of OSI);

(d)   WG 4: OSI management;

(e)   WG 5: specific application services;

(f)   WG 6: Session and Presentation Layers and common application services.

The international consensus sought by ISO takes a long time to achieve. Working meetings cannot take place more frequently than 6 to 9 months. It takes that length of time for the results of a meeting to be typed, distributed internationally, reviewed by national member bodies (which can require several meetings in each nation), to be commented upon and those comments distributed in advance of the next meeting. Only by this process can it be ensured that full national consensus is taken to each international meeting by delegates and these in turn can be synthesised into an international agreement.

It can take many meetings to arrive at a document which is accepted by all the participants. At that time the formal voting procedures are started.

A potential International Standard (IS) must first pass through the stages of draft proposal (DP) and draft international standard (DIS) before becoming an IS. At each stage there is a vote by member bodies of ISO as to whether they find the document acceptable.

A document which has the acceptance of the participants working on it is first balloted as a DP. The ballot period is normally twelve weeks and the votes can be simple 'Yes', 'Yes' with comments or 'No' with comments. After balloting, an attempt is made to reduce the number of negative votes and resolve all the issues raised by the comments received with the votes. It is permissible for a small number of negative votes to be unresolved at the time of progression to DIS. A similar balloting procedure is performed, with the same rules, before the document is deemed ready to progress to IS. The emphasis at all stages of balloting is to achieve *substantial* support, not just a simple majority. At that time it is sent to the Central Secretariat of ISO for processing to become a published IS. The processing can involve minor corrections of an editorial nature, production of French and Russian translations (these being the two official ISO languages besides English) and typesetting, and can take around one year to complete.

2.3   INTERNATIONAL TELEPHONE AND TELEGRAPH CONSULTATIVE COMMITTEE (CCITT)

CCITT is an international organisation whose aim is the production of standards for international interworking of telephone and other telecommunications systems. Conformance to its standards is a mandatory requirement on its full members.

Full members of CCITT are normally the PTT (telecommunications carrier) of each nation. Observer membership is open to any other organisation. Observers may participate in meetings but may not vote.

The objectives of CCITT result in different working procedures to those adopted by ISO. CCITT has adopted a four-year work cycle. At the end of each four-year cycle, CCITT recommendations are all republished together in a series of volumes and each is informally known by the colour of its binding (the 1984 volumes were published in red bindings and called the 'red book').

Because of the need to stop work and publish after four years, CCITT recommendations are often published unfinished. Areas which are incomplete or unstable are denoted 'for further study'. However, the need for international telecommunications to be developed in a harmonious and compatible way and the recognition that PTTs have investment programmes to adhere to makes publication of such unfinished recommendations preferable to waiting until all the smaller details are finalised.

Determination of what will be published is made by voting which takes place at the end of each four-year study period (see below).

CCITT divides its area of interest between a number of study groups (SGs), each being denoted by roman numerals. The study groups concerned with areas related to OSI are SG VII, SG VIII and SG XVIII.

SG VII is titled 'Data Communications Networks' and covers facilities (e.g. reverse charging, call redirect, etc.), interfaces (e.g. X.25, X.21, etc.), interworking, maintenance and message handling (X.400). As the major providers of networks, PTTs are clearly concerned with layers 1 to 3 of the OSI Reference Model. As providers of telematic and message services they are also concerned with the higher layers.

SG VIII is titled 'Telematic Services' and is concerned with facsimile, teletex, videotex and character sets.

SG XVIII is titled 'Digital Networks' and is currently producing standards for all aspects of the Integrated Services Digital Networks (ISDN).

The work of each SG is subdivided into 'questions'. Each question is a defined topic of work and these can range between one question covering the whole of the OSI Reference Model (SG VII, Question 42, denoted Q42/ SG VII) or a question on updating a published recommendation. Questions which are related can be grouped together and allocated to sub-working parties to pursue.

CCITT study groups hold a 'plenary' meeting at the end of each four-year study period at which member bodies vote on the text produced in answer to each question. All approved texts are included in the new volumes of recommendations.

The constitution and procedures of CCITT result in recommendations which are orientated specifically to the needs of PTTs. It is up to each PTT to make its own arrangements to try to ensure that the services agreed upon in CCITT meetings will satisfy user demands in their own country.

2.4 EUROPEAN COMPUTER MANUFACTURERS' ASSOCIATION

As its name indicates, ECMA is an organisation of computer manufacturers. Full membership is open only to companies who develop, manufacture and sell computers in Europe. Associate membership is available to companies who satisfy only some of the full set of criteria.

As with CCITT, the restricted membership makes full consensus amongst participants in standards-making easier and quicker to reach than in ISO. Coupled with simple secretariat procedures, this allows ECMA to produce standards much quicker than ISO and, in fact, quicker than CCITT.

For items of computer equipment such as magnetic-tape drives, where it is necessary to define the encoding technique to be used on the tape so that it can be interchanged between drives manufactured by different companies, the constitution and procedures of ECMA are quite sufficient. However, OSI spans a much wider community of interest. Although manufacturers must be willing and able to implement OSI, CCITT members must also supply the services of the lower layers for wide area national networks and the end-users must be willing and able to make use of the resulting totality of services for their specific applications. Thus, the interests served by ECMA are only a part of the totality of parties interested in OSI.

ECMA divides its work amongst a number of technical committees (TCs). The TC concerned with OSI is TC 32.

TC 32 is titled 'Communication, Networks and Systems Interconnection'. This committee covers all the layers of the OSI Reference Model.

In addition, TC 29 is titled 'Text Preparation and Interchange' and is concerned with broadly the same area as CCITT SG VIII and ISO/TC 97/SC 18.

2.5 LIAISON AND CO-OPERATION

As indicated above, the community of interest in OSI is very wide and the perceived need within that community is for international standards. ISO is the only body capable of producing the standards needed but as both CCITT and ECMA have interests in OSI and can publish their own standards, a great deal of co-operation is needed between all three bodies.

The required co-operation happens in a number of ways.

Firstly, both CCITT and ECMA are liaison (L) members of ISO and this means that ISO can generate and send 'liaison statements' to them officially requesting comment or drawing attention to some area of concern. Also, each organisation can nominate official 'liaison representatives' to attend meetings of the other organisations.

Finally, it often occurs that there are a number of delegates common to two or more organisations who normally attend each set of meetings.

In these ways, co-ordination between the efforts of different organisations can be achieved.

There have, however, been a number of problems with the co-ordination of OSI standards between CCITT and ISO which have required special additional procedures. It was agreed very early in the OSI standardisation effort that there should be identical wording of OSI standards when they were to be published by more than one organisation. With ECMA this posed no special problem because in cases where they develop their own standards early, ECMA will normally replace them with the ISO text when it reaches stability.

The four-year study cycle of CCITT causes problems, however, because any publication of a document by CCITT before the ISO document has become stable, even if it were identical to the ISO text at that time, would remain current in the CCITT publication for four years, and often significant technical changes take place in an ISO draft before it is published as a full International Standard.

The mandatory nature of CCITT recommendations and consequent careful differentiation between 'essential' and 'additional' services (of the Network Layer in particular) have introduced an unusual number of 'political' problems into the ISO discussions.

In order to resolve these problems, ISO have tried to schedule its activities to accommodate the 1984 CCITT publication date and made special efforts to achieve stability of both Session and Transport Layer standards in time for the 1984 CCITT Plenary meetings.

In order to arrive at mutually acceptable texts, CCITT and ISO have held joint meetings. In strict terms this is not allowed by the CCITT rules but meetings have been arranged to take place concurrently and at the same venue so that a free interchange of delegates between CCITT and ISO could occur.

The result has been texts which are identical except for those minor variations in wording necessitated by the rules of the two organisations, particularly the addition of 'considerata' at the start of the CCITT recommendations.

CHAPTER 3

The Reference Model

3.1 LAYER PRINCIPLES

The OSI Reference Model ISO 7498 was developed as a basic framework to deal with the complexity of communicating systems and the required protocol standards.

In the same way as the principles of structured programming have been accepted as essential for dealing with large-scale complex software projects, it was realised that some analogous methodology was required for dealing with the complexity of communicating computer systems. In fact, many of E. W. Dijkstra's remarks in *Structured Programming* are very relevant, particularly one of his conclusions:

'We should recognise that already now programming is much more than an intellectual challenge; the art of programming is the art of organising complexity, of mastering multitude and avoiding its bastard chaos as effectively as possible.'

The chaos in the data communications was already beginning to show in the mid 1970s. There was no overall framework for the co-ordination of the development of data communications protocol standards, nor for placing the existing standards in perspective.

From analysis of typical communicating computer systems it became clear that the multitude of functions to be performed, and the resulting protocol elements, necessitated some methodology and structure for organisation into manageable parts.

A 7-layer basic architectural structure was chosen. This structure is an abstraction of the real system, enabling freedom of the internal implementation within the real system, whilst specifying the requirements to be met for the externally visible behaviour (the protocol) of the real system which, of course, is the vital element when connecting remote systems together. The protocol operating between equivalent entities at the same level but resident in different remote systems is known as the peer-to-peer protocol.

Detailed descriptions of each individual layer will be given in later sections. This section will only deal with basic architectural principles, how the magic number of seven came to be chosen as the number of layers, the factors that were taken into account in deciding the placement of boundaries.

Ten guiding principles were drawn-up. These are listed in the Reference Model and are summarised as follows:

P1: keep the overall structure simple;

P2: create boundaries only at points where interactions across the boundaries are minimised, (i.e. keep linkages between modules as simple as possible);

P3: separate into different layers those functions that are very different in nature and purpose;

P4: collect together within a layer those functions that are similar or highly inter-related;

P5: select boundaries at points which past experience has shown to be successful;

P6: create layers such that the internal mechanisation of such layers can be made independently from the functionality that it provides (for flexibility of protocol, and evolution);

P7: create boundaries where it may be useful at some time to have a real physical boundary (for a front-end processor, or distributed processor system);

P8: create a layer where there is a need for a different level of abstraction in the handling of data;

P9: create layers such that changes of functions and protocols can be made within a layer without affecting other layers. (This appears to be the same as P6);

P10: create for each layer boundaries with its upper and lower layer only.

The discussion below illustrates some of the reasons why application of these principles has led to the OSI architecture.

Computer systems provide a specific end-product (or products) for their users and purchasers. For example, the end-product may be an airline seat reservations system, a distributed order entry or inventory control system, or perhaps a general-purpose time-sharing bureau service. Generically, these end-products are known as the applications.

Applications will clearly be many and various. OSI must not only accommodate existing applications, but those foreseen in the near future, and those as yet unforeseen but which will inevitably be introduced well into the future.

A vital component between remote communicating systems is some form of telecommunications medium, e.g. single leased line, leased line network, public switched network, physical medium, etc. These are known generically as real networks. In the same way as applications are many and various and rapidly evolving so, too, are real network technologies e.g. packet networks, local area networks, satellite networks, cable-TV networks, etc.

It is clear that this rapid development of application and network types should not be inhibited. Moreover, it has to be possible to operate any given application over any given real network. It is not acceptable to redesign applications every time a new network technology is encountered. Further, it is a requirement to be able to interconnect real networks together and operate a given application over any combination of different real network technologies.

The above considerations led to the fundamental division of the model of the architecture of a communicating system architecture to provide a stable functional boundary over which any application can operate irrespective of any real network technology. This fundamental division occurs at the Network Layer of the Reference Model. In addition to being real-network independent, the Network Layer has to be able to organise the interconnection of many different real networks in such a way as to provide the appearance of a single network to the architecture above. It is important to realise that this difference between this 'idealised network' and real networks is denoted in the Reference Model by use of capitalised letters. Thus the OSI boundary is the Network Layer and the functionality it provides is known as the Network service. The use of the word 'Network' (with capital N) in these cases is logically distinct from the word network (small n) as applied to particular technologies (e.g. an X25-based packet switched 'network', local area network, etc. (see Chapter 4 for further refinement of these concepts).

However, even given this stable functional boundary some very specialised applications may have requirements that it would be unreasonable to expect all networks to provide. For example, extremely low error rates, or high degree of security may be important. Where end-to-end assurance is important it can only sensibly be provided directly between the remote communicating systems, themselves. That is, independently of network node and/or gateway operations. Additionally tariff optimisation of a given network can only be performed above the level of the network itself. There may be other 'quality' requirements that applications have which may not be adequately provided by real networks, (e.g. a wide range of throughput capabilities, priorities, etc.).

Thus, while applications need a stable functional boundary they also need one which will provide the required Quality of Service (QoS) appropriate to the application, and in a cost-optimised fashion. This leads to the formation of the Transport Layer which provides the same type of service as the Network Layer, but which is able to bridge the QoS gap between that required by the Application and that available from the Network Layer.

The diversity of functions above the Transport Layer and of technologies below the Network Layer gives rise to the 'hour-glass' or 'wine glass' picture sometimes used in representations of the Reference Model (see Fig. 3.1).

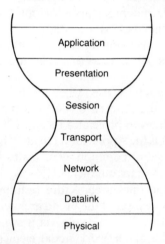

Fig. 3.1 The hour-glass model.

The architecture below the Network Layer typically reflects architectures recently developed for public switched data networks. It has become accepted that such architectures should contain three separate layers, one for the physical and electrical characteristics of the interface to the network, one for the data transmission procedures to transfer information to the data switching exchange, and one for passing control and data information relating to the switching functions, (i.e. to the call-processing function within the exchange). The first two layers are designated Physical and Data Link Layers respectively, and the latter belongs to the Network Layer. This is an application of principles P3, P5 and P7. Additionally, P8 particularly applies to the Data Link Layer operating above the Physical Layer.

A study of applications reveals that many applications have a number of common requirements. For example, applications may need to precisely control the interactions between themselves, reflecting the end-user operations in some cases. There is a need to control which of the two applications is permitted to transmit at a given time, and then subsequently pass permission to the other application. Applications may also require a synchronisation service to accommodate the kind of substantial data losses associated with system-type failures in end-systems, e.g. disk failure, etc. (only minor data losses incurred during line transmission or circuit failure can be recovered autonomously by the lower layers, usually the Data Link Layer and Transport Layer respectively). The Session Layer of the Reference Model is the layer created for assisting with provision of these functions. Recovery may be at the level of granularity of files, disk sectors, pages, documents, etc. as appropriate to the application. Principles P3 and P4 were applied in producing the Session Layer.

Another item of extreme importance to applications is the format (i.e. encoding) in which the data is transferred. Furthermore, it may be necessary to introduce conversion functions whilst preserving the semantics in cases where the encodings of one system are different from that of another. In any event there is a requirement for identifying and possibly negotiating the encodings to be used when information is transferred. The Presentation Layer of the Reference Model is the layer created for this purpose, and again relates to principles P3 and P4.

The final upper layer is the Application Layer, containing the application itself and including a set of common elements to be used by families of applications with common requirements.

Figure 3.2 shows the final architecture resulting from the foregoing considerations and application of the broad architectural principles. In the case where systems are communicating via a switched data network of some kind, the 'intelligent' network nodal elements are represented as intermediate systems as shown in Fig. 3.3. There may be many such intermediate systems interposed between the end-systems, in the case where

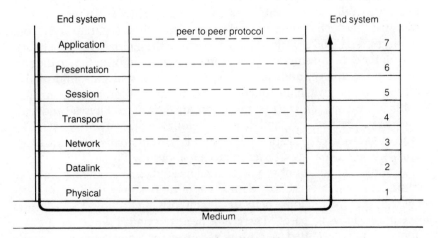

Fig. 3.2 Final OSI architecture.

the path of the communication has been routed through many switching nodes.

It is possible to further subdivide individual layers into sublayers (this will be seen later in the cases of the Network and Application Layers).

The main ten basic principles can be re-applied to sublayering.

In addition, three more principles are given for sublayering:

P11: create further subgrouping and organisation of functions to form sublayers in cases where communications services need it;

P12: create, where needed, two or more sublayers with a common and therefore minimal functionality to allow interface operation with adjacent layers;

P13: allow by-passing of sublayers.

P13 in particular indicates the main difference between a layer and a sublayer.

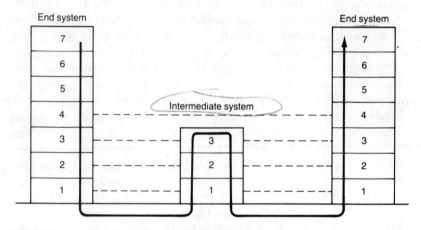

Fig. 3.3 Intermediate systems.

The OSI Reference Model specifies one particular layered architecture. Many other proprietary layered architectures exist, e.g. IBMs, SNA, DECNET, etc. These may result from the same basic layer principles but may have located the actual layer boundaries in places different from the OSI Reference Model. Some may even have seven layers (but organised quite differently), and the functionalities quite different in part or in total. Claims for layered architectures need to be carefully evaluated as to their relationship to the OSI Reference Model. However, layering itself is useful for assessing the potential of a manufacturer's architecture for accommodating OSI protocol standards. It is easier to modify systems if they have a modular structure even if it differs from OSI in some ways.

Some system developers are introducing seven layer boundaries in order to accommodate migration from non or only partial OSI compatibility to the full OSI set of protocol Standards in the future.

Purchasers of systems are well advised to consider these issues when assessing manufacturers' proposals during their contract evaluation exercise.

3.2   THE CONCEPT OF A LAYER

Having made the layer the fundamental building brick in the Reference Model it is now necessary to define it and its capabilities more rigorously. This will only be done in general terms by introduction of a 'shopping list' of functions and capabilities. Details for particular individual layers will then draw from this set as appropriate, when the particular layer is discussed. It should be noted that the OSI Reference Model itself does contain sections for individual layers. This is unfortunate in the author's view, and has led to unnecessary conflict between the architects and layer experts and associated standards. It is clear that beyond a certain granularity, details about a given layer should be left to the experts developing the particular layer. With hindsight, it would have been better if the Reference Model had only dealt with the general architectural principles and not included detailed layer descriptions which would inevitably be superseded, as detailed work on individual layers progressed.

The urgency to publish the OSI Reference Model itself before all the layers were completed in detail has resulted in divergence between some layer descriptions in the OSI Reference Model, and the individual layer standards being progressed. For this reason, this book will ignore the Reference Model layer description (section 7 of RM) and will instead use the definitive work of the individual layer groups.

From now on, it will be necessary to use terms more precisely and thus it becomes necessary to generate a multitude of definitions. We will try, however, to build them up in a logical fashion.

It is clear that there is a strict hierarchical relationship of the layers. It thus follows for any given layer there is a knowledge of: (a) what exists below and (b) what the next layer above expects. There is a lot of misunderstanding about layer independence. Total layer independence is clearly not possible in such a hierarchical ordering of functionality. Only certain aspects are independent as will be described when encountered.

Generally, then, when considering a specific Layer, say Layer (N), it is necessary to consider its relationship with the adjacent layers, (N–1) and (N+1). We thus formally speak about the (N)–Layer, the (N+1)–Layer and the (N–1) Layer. When referring to specific layers by name the (N)–,

(N+1)-, (N-1)- prefixes are replaced by names of the layers, e.g. Transport Layer, Session Layer and Network Layer, etc. It may also be said that the (N+1)-Layer is the (N)- Service-user.

To simply and abstractly define a layer the concept of a layer service is introduced. This permits an implementation-independent description to be achieved for a layer in terms of its capabilities. Thus, the functional requirements may be produced in a top-down fashion leaving the related protocols to be developed from the requirements specification.

The (N)-Service is defined as: the capability of the (N)-Layer and the layers beneath it, i.e. the combined functionality of all layers up to and including the (N)-Layer, and available to Layer N+1.

Thus, for each layer of the Reference Model there will be a standard defining the layer service, designated the Layer service definition, and one (or more) standards specifying a protocol which fulfils the requirements of the service definition. The latter are designated protocol specifications.

3.2.1 Modes of operation

Before proceeding further it becomes necessary to explain two different kinds of service, the connection-orientated mode (COM) and the connectionless mode (CLM).

Originally, the Reference Model only considered connection-orientated mode. The connection-orientated mode is based on classical sign-on work, sign-off modes of operation. So, for example, the kind of application would be one of reasonable duration, several minutes of longer, involving perhaps a log-in, log-out procedure at the user-computer level, with perhaps a circuit switched connection (real or virtual) between the communicating parties. For this type of operation the associated (N)-Layer would clearly have a connection-establishment phase, a data transfer phase and a disconnection phase. Some 'long-term' association is established for the duration of the data transfer phase, between two communicating parties. This association is created at establishment time and destroyed at disconnection time.

For connection mode operation, an (N)-connection is defined as 'an association established by the (N)-Layer between two or more (N+1)-entities for the transfer of data'.

Connectionless mode, on the other hand, does not have three phases of operation. There is no connection establishment or disconnection phase, and no negotiation of the association between communicating parties. Every time a unit of data (called the Unitdata) is exchanged it is totally self-contained, and thus every piece of data has all the necessary control information associated with it, e.g. the calling and called addresses and any other typical connectionless mode operations. An addendum to the

Reference Model has been produced for the connectionless mode of operation, the details of which will be discussed later.

Many of the architectural principles and elements are applicable to both modes of operation, whilst some may only be applicable to the particular mode. Generally, the principles and associated elements are derived from the basic concept of a layer.

The functions in a layer, sometimes collected into groups, are performed by elements called 'entities'. Entities are capable of communicating with other entities in the same layer of different systems using one or more protocols. An entity will provide the functions so that the layer above can use just one or either mode of communication. In other words, an entity could just contain the functions to provide a connection-orientated mode of communication, or it could also contain the functions to support a connectionless mode of operation (in which case it provides both connection-orientated and connectionless modes of communication), or it could just contain the functions to provide a connectionless mode of communication.

If more than one protocol is available in a particular layer then the choice of protocol is a function of that layer and not the layer above which is using the provided service.

3.2.2 Service access points

Operations between layers are said to occur at the Service Access Point (SAP) for a given instance of communication, i.e. for a given connection (out of the set of all connections) or for a particular Unitdata transfer (in the case of connectionless mode operation). For a given layer (N) then the (N)–SAPs are located on the (N)–Service boundary, which is the boundary between the (N)–Layer and the (N+1)–Layer. So, for example, the network SAPs are located between the Network Layer and the Transport Layer. Thus, it is said that an NSAP identifies a Transport entity.

Associated with each SAP is an address. Only one entity is allowed to reside above a particular SAP and thus the address of that SAP can be used to identify the entity. The same addressing mechanism is used for connectionless communication and during the connection establishment phase of a connection orientated communication. Once an (N)–connection has been established between two (N)–SAPs, and, hence, between two (N+1)–entities, then any information which is passed on to the connection through one SAP will leave the connection through the other SAP and hence no further addressing is required. If the SAP has more than one connection associated with it, then the required connection is indicated by using a 'connection end-point identifier' which is similar, in principle, to a logical channel number in X.25. General address principles and layer specific addressing mechanisms will be described in later sections.

Not all of the operations between layers are defined. If a particular inter-layer exchange does not result in a significant change in the external behaviour of the system then it is considered a matter local to the system and, hence, not a proper subject for OSI standardisation. Examples of this will be detailed later when encountered.

The basic concept of a layer has now been described and is summarised in Fig. 3.4, which shows all the items described so far. The figure is arranged to show the (N)–Layer spanning two remote (N)–Layer entities communicating via the peer protocol between them. The combination of the two entities is necessary to show that the overall service of a given layer is dependent on both communicating systems. This will become clear when the principles of service definitions are dealt with in more detail.

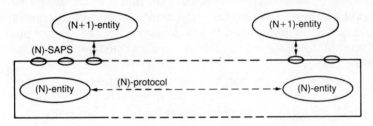

Fig. 3.4 The (N)-layer.

For connection mode operation, the (N)–connection (defined as 'an association established by the (N)–Layer between two or more (N+1)–entities for the transfer of data) is graphically illustrated as in Fig. 3.5.

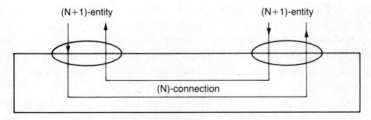

Fig. 3.5 An (N)-connection.

3.3 LAYER SERVICE PRINCIPLES

It has already been observed in Chapter 1 that the OSI Standards must be widely applicable and in particular must be implementable on any computer architecture within any operating system structure. It is therefore important that the terminology used within such Standards is not specific to *any* style of implementation.

It is particularly important not to imply any specific means of binding the elements of layer protocol to protocol in adjacent layers. Nevertheless, it has been necessary to specify the way in which a protocol is carried by the lower-layer protocol and the effects of layer events on higher-layer protocol.

The 'service definition' for a layer specifies in an abstract, implementation independent fashion a conceptual set of 'events' that can be passed from one layer to the next.

Many implementations of 'services' are possible. It may be that when an Application has data to transmit, a library procedure is called to perform Presentation Layer functions which in turn calls a procedure to perform Session Layer functions. Alternatively, each layer may be implemented as a separate 'task' within a multi-tasking system or even as separate pieces of hardware. A combination of each of the above and/or other techniques could be used.

The important thing is that, however it is implemented within a system, that system should exhibit the same behaviour to the outside world as that which is described by the combination of layer standards which have been implemented.

The abstract elements defined in the service definition are aimed at describing what must be regarded as integral units of information of various types whose integrity must be maintained on an end-to-end basis between the two communicating service users.

The sequences of events between adjacent layers that occur via the service access point is described in the service definitions by a set of named service primitives. Each service primitive is defined in the form similar to that of a subroutine or procedure call in a computer program, i.e. it has a specific name and associated input or output parameters. The sequence of permitted service primitives is defined by the use of state or time-sequence diagrams. State diagrams are used to describe local events within a layer and are used to show the relationships between events across the local layer service boundary, and the consequent events within the layer. On the other hand, time sequence diagrams are used to describe the relationship of events across the service boundary at one end of a communication and the concommitant events across the service boundary at the remote end.

The conventions to be used for service definitions are contained in a technical report, TR 8509, and are summarised below:

3.3.1 Service primitives

Primitives for connection-mode operation are classified into four types as follows:

(Name) Request

This is initiated by the (N+1)-Layer (i.e. the user of the (N)-service) to request or activate a particular *named* service from the (N)-Layer.

In order to set up a network connection, the Transport Layer would issue an N-Connect Request, for example.

(Name) Indication

This is issued by the (N)-Layer (i.e. the service provider) to (N+1)-Layer (the service user), to advise of the occurrence of the *named* service event.

To continue with the example, a system receiving a request for a network connection would issue an (N)-Connect indication to the Transport Layer which can then subsequently decide whether to accept or refuse the request for a network connection.

(Name) Response

This is used by the (N+1)-Layer in reply to the received *named* indication from the (N)-Layer.

In our example, a N-Connect Response from the Transport Layer would inform the Network Layer in the receiving system of the willingness to accept the incoming network connection.

(Name) Confirm

This is issued by the (N)-Layer to indicate to the originator of the Request for the *named* service that the service has been completed.

For our example, an N-Connect Confirm from the Network Layer would inform the Transport Layer in the originating system that a network connection had been successfully established.

3.3.2 Time sequences

The time sequence in which primitive events can take place has to be specified and is illustrated by time-sequence diagrams in the service definition standards.

The time sequence diagram relating to our example above is shown in Fig. 3.6. Time increases down the vertical time lines.

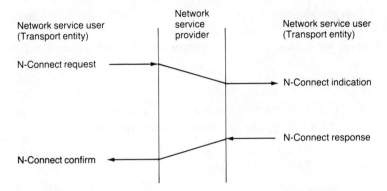

Fig. 3.6 Confirmed service time sequence diagram.

The relative positions of the arrows representing the primitives is of the utmost importance. From the figure the position of the Indication means that this event can only take place *after* the corresponding Request has taken place. Similarly, in this case the Confirm can only occur provided the Response has already occurred. At first glance this all seems obvious. However, there are other sequences which can occur for other types of service which need to be distinguished.

A different type of sequence can be illustrated by a particular type of disconnection sequence. In our first example there was a complete end-to-end service user to service user (via the provider) handshake. This is known as a confirmed-service-element. Some disconnection procedures are not like this, since the initiator will wish to break a connection irrespective of whether the other service user agrees or not. In this case, the time sequence diagram will look like Fig. 3.7. The use of the tilde (˜) in this figure means that the occurrence of the Confirm is not related to the time at which the Response is issued at the remote end, nor even the time at which the indication occurred at the remote end. All one can say is that the confirm can occur at any time after the request irrespective of events at the remote end. This type of service is known as an unconfirmed-service-element.

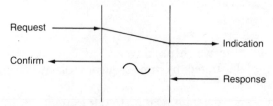

Fig. 3.7 Unconfirmed service time sequence diagram.

Only two primitives are necessary for connectionless-mode operation, and the resulting time sequence diagram is shown in Fig. 3.8.

It can be seen that there is no way that the sender of an (N)-Unitdata Request can have knowledge that the equivalent (N)-Unitdata Indication has occurred at the remote end. The implications of this will be discussed later but generally it means that confirmation of delivery has to be dealt with (if required) at some higher layer.

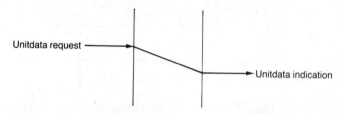

Fig. 3.8 Connectionless service time sequence diagram.

3.3.3 Service units, interface units and protocol units

As stressed previously it is important to distinguish between the abstract nature of the layer service boundary and any local implementation of such a boundary in the form of an interface. The protocol standards meet the requirements of the abstract definition. Thus in any real implementation the protocol may in fact be mapped to different local interpretations provided that the same semantics are preserved.

The prime requirement is to ensure that each of the communicating systems has the same understanding of the transmitted/received protocol elements irrespective of differences in two local system implementations.

Within the layer itself different types of protocol elements will be exchanged between the peer entities. These types and their associated requirements (sequence numbers, parameters, etc.) are protocol data units (PDU), which comprise the 'within' layer protocol control information (PCI) together with data (if any) from the higher layer (see Fig. 3.9).

A unit of data from the higher layer whose integrity is to be maintained end-to-end between the two service users is termed a service data unit (SDU). The length of this data unit is not fixed (except in certain cases) and for the general case is arbitrarily long and coincides with the sizes that the applications find meaningful (e.g. a record, a transaction unit, etc.). This leads to the possibility of three different-sized units being identified, one for internal system operation and one for protocol operation and one for the application. The sizes used for protocol operations are optimised around transmission error rates, propagation delays, retransmission strategies, etc.,

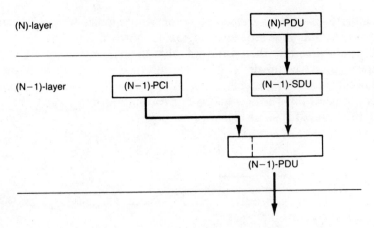

Fig. 3.9 Protocol control information.

considerations which do not have to be taken into account internally. We can thus distinguish between the service data units operating at a conceptual level, for application purposes, the interface data units operating across a local service boundary, and the protocol data units operating between systems.

Figure 3.10 shows an amount of (N)-layer data comprising a complete single (N–1)-SDU passing to the lower layer via a series of interface data units (IDUs), according to local system implementation. These IDUs are

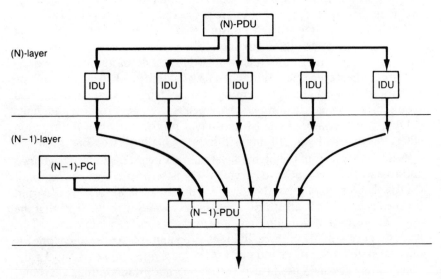

Fig. 3.10 Interface date units.

then combined to form the (N–1)-PDU by addition of the appropriate (N–1) protocol-control-information (PCI). This figure shows only a simple case where the SDU is contained within a single PDU.

In general, it will not be the case that one (and only one) SDU is accommodated within one (and only one) PDU. There may be a many-to-one or one-to-many mapping as well, and these cases have formal definitions for the mechanisms involved. This situation arises from the fact that an SDU is of arbitrary size according to the particular application. Applications of certain types may only transfer a small amount of information at a time, others may transfer much larger amounts. The protocols themselves will have PDU sizes arranged in accordance with the kind of protocol in operation. For example, one would expect a Data Link Layer protocol to employ frame sizes optimised to the transmission error rate and recovery procedures. Such sizes may be quite different from those being handled by the application as a meaningful unit for processing.

The case where several PDUs are required for one SDU is termed 'segmentation' (for the transmitter) and 'reassembly' (for the receiver). Figure 3.11 shows an SDU segmented over two PDUs. There could, of course, be as many PDUs as necessary dependent upon the relative sizes. As far as the protocols are concerned, protocols which provide this feature contain a mechanism for 'chaining' the SDU data across the PDUs, usually by a sequence numbering and last/not last flag scheme.

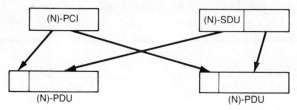

Fig. 3.11 Segmenting.

Figure 3.12 shows blocking and deblocking (again related to transmitter and receiver respectively). The process of dealing with interface data units is omitted here for simplicity and since it is a local system implementation

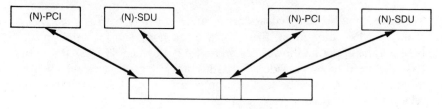

Fig. 3.12 Blocking and deblocking.

detail. The precise way in which the integrity of an SDU is maintained is a protocol issue and may include extra PCI information within the PDU.

The last item to be covered related to service boundary units/protocol operations is concatenation and separation. In the previous examples there was data crossing the boundary. However, within a given layer there will be PDUs which operate solely on a peer-to-peer basis within the layer, and do not contain data from the higher layers. These PDUs, of course, may also be combined with others, including those containing higher layer data for transfer as a single SDU of the lower layer. Figure 3.13 illustrates this case.

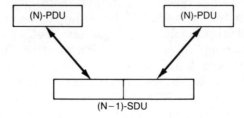

Fig. 3.13 Concatenation/separation.

We can see by now that the protocols of one layer are nested inside the protocols of the lower layer. Thus, in general the SDU of a given layer is treated transparently by that layer but will, in fact, contain not just data of the upper layer but upper layer protocol as well. So if one were to monitor the path between two remote end-systems for all layers in operation simultaneously one would see the nested structure as shown in Fig. 3.14. For simplicity just PDUs with a simple PCI and higher layer data are shown.

Link PCI	Network PCI	Transport PCI	Session PCI	Application PCI + data encoded in transfer syntax	Link FCS

Fig. 3.14 Nested PDUs.

3.4   GENERAL ADDRESSING PRINCIPLES

We have already seen that addresses are associated with service access points (SAPs). In general, then, for a given layer (N) the relationships between the upper and lower boundary SAP, i.e. the (N+1)-SAP and the (N)-SAP need to be considered. At this stage the full generality will be discussed. Later, it will be seen that only particular subsets of all possible relationships are permitted for certain layers.

There are four possible cases depending on the relationships between the addresses and how the binding between (N)-SAPs and (N–1)-SAPs is achieved.

The simplest case is where there is a one-to-one fixed binding between (N–1)-SAPs and (N)-SAPs as shown in Fig. 3.15(a). In this case the (N–1)-SAP serves to identify the (N)-SAP as well without any further addressing being required.

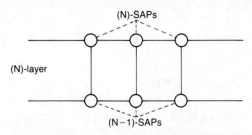

Fig. 3.15(a) One-to-one address mapping.

Figure 3.15(b) shows a one-to-many correspondence between (N–1)-SAPs and (N)-SAPs. This can be further divided into two distinct subcases depending on whether there is a strict hierarchical relationship between (N–1)-addresses and (N)-addresses. A hierarchical address scheme is one in which the (N)-address is always mapped into only one (N–1)-address. So, for example, in Fig. 3.15(b) (N)-addresses X, V and Z are only ever associated with (N–1)-address A, and similarly P, Q and R and only ever associated with B. Thus, the (N+1) entities are defined by the combination of the (N–1)-address and the (N) address, say AX, for example. This means that to identify unambiguously the (N+1)-entity X does not itself have to be unique provided that A is unique. In this case, it is usual to call the value X the (N)-suffix.

If, on the other hand, an (N)-address can be mapped into several (N–1) addresses, or an (N)-address is not permanently mapped into the same (N–1)-address then the address scheme is said to be non-hierarchical. The consequence of this is that both the (N–1)-address and the (N)-address need to be uniquely identifiable. In this case, however, each (N)-SAP can be

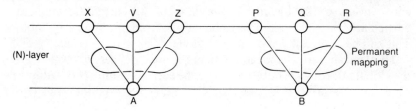

Fig. 3.15(b) Many-to-one hierarchical address mapping.

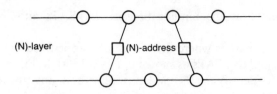

Fig. 3.15(c) Non-hierarchical address mapping.

addressed independently from the (N–1)-SAP. Figure 3.15(c) shows the non-hierarchical case.

The term 'unique' as used in this context needs to be further explained. It applies to the value of the address and not necessarily to the service access point itself. For example, it is not the case that an (N)-entity is uniquely identified by the (N–1)-address since we have seen that an (N)-entity may be connected to more than one (N–1)-SAP. So there may be several addresses (or synonyms) for a given (N)-entity. However, each individual address is unique in the sense of being unambiguous from all other address allocated. As an analogy, let us suppose that the authors each have two telephones. Clearly two different numbers can identify K. G. Knightson and two more can identify T. Knowles. However, it is obvious that the numbers for K. G. Knightson must also be different from those for T. Knowles (i.e. the numbers must be globally unique). However, it cannot be said that K. G. Knightson and/or T. Knowles can be uniquely identified by a single global number.

The above properties may be summarised formally as follows:

(a) an (N+1)-entity may concurrently be attached to one or more (N)-service access points attached to the same or different (N)-entities;

(b) an (N)-entity may concurrently be attached to one or more (N+1)-entities through (N)-service access points;

(c) an (N)-service access point is attached to only one (N)-entity and to only one (N+)-entity at a time.

Generally it is assumed that addresses identify places where the service access points can be found. However, there may be cases where a given entity needs to be identified regardless of its current location, and particularly when it is required to be mobile from one system to another. This can be covered by allocation of global titles to (N)-entities.

This leads in turn to the requirement for directory services which, given the global titles of (N)-entities can yield the corresponding (N–1)-addresses. Thus, an (N)-directory translates global titles of (N)-entities into (N–1)-addresses.

Routing is also defined as a function which translates the (N)-address of an (N+1)-entity into a path or route by which the (N+1)-entity may be reached. This will not be discussed further here, since routing is constrained to be only available and applicable to the Network Layer and thus will be dealt with in detail in section 4.3.4.

There is no restriction in principle to the number of connections that can be established to a given SAP, which may only have one address. Now the protocols which enable this to be achieved (take the CCITT Recommendation X.25 on packet switching, for example) must contain adequate identifer mechanisms. Conceptually, this means that at the service access point there has to be an (N)-connection endpoint identifier for every (N)-connection within the same (N)-service access point. In reality, whether these exist separately from the identifiers within the protocol, and how they are allocated within and/or between service access points is a local implementation issue. Additionally, this means that absolute values for connection endpoint identifiers are not passed end-to-end, nor can any relationship between those used in the two remote systems be assumed or deduced.

Finally, it should be noted that, in general, there is no restriction on the number of addresses associated with a service access point and vice versa. A given service access point may have one or more addresses and conversely, a number of service access points may have the same address.

3.5 RELATIONS BETWEEN LAYER CONNECTIONS

Figure 3.5 showed two remote (N+1)-entities communicating via an (N)-connection.

In fact, three possible relationships between (N+1) and (N)-connections can exist. Figure 3.16(a) shows a simple one-to-one relationship where only one (N+1)-connection exists concurrently over one (N)-connection. Also, for simplicity only one (N)-connection is shown between the (N)-SAPs, but in general there could be more than one as discussed towards the end of the

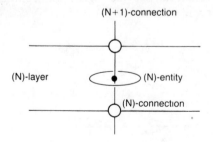

Fig. 3.16(a) Simple one-to-one connections.

previous section. The prime purpose of the figure is to show the relationship between (N)-connections and (N+1)-connections.

3.5.1 Multiplexing

Another possibility is the establishment of more than one (N+1)-connection over an (N)-connection. This is formally known as multiplexing (for the transmitter) and demultiplexing (for the receiver). This is illustrated in Fig. 3.16(b) which shows three (N+1)-connections established over 1 (N)-connection. From a practical standpoint this technique is used to optimise the use of the bandwidth available from the (N)-Layer (in the case where it is greater than that required for any individual (N+1)-connection) and is confined to the Transport, Network and Data Link Layers.

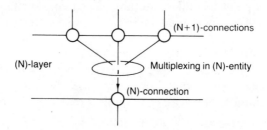

Fig. 3.16(b) Multiplexing in the (N)-entity.

3.5.2 Splitting

The remaining possibility is the establishment of one (N+1)-connection over several (N)-connections. This is formally known as splitting (for the transmitter) and recombing (for the receiver). Figure 3.16(c) illustrates such operation over three (N)-connections. In practice, this is used where the bandwidth requirement of the (N+1)-connection is far greater than any individual (N)-connection can provide. Again, use of splitting is confined to the Transport Layer and below.

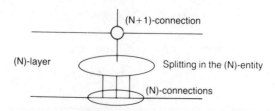

Fig. 3.16(c) Splitting in the (N)-entity.

It should be noted that these relationships are orthogonal to those of SAPs. For example, in Fig. 3.16(c) the relationships between connections at Layer (N) and (N+1) is independent of whether there are three (N)-SAPs or one (N)-SAP with one (N)-connection each. Similarly, there may be one or three addresses.

3.5.3 Multi-end-point connections

The most common type of connection is that established between a pair of remote SAPs. However, there are cases in which multiple associations between entities can occur. These can still be 'modelled' as multiple instances of n-pairwise connections.

A centralised multi-end-point point is one in which the connections all terminate at one system. Thus data transmitted by the entity at the centralised multi-connection endpoint is received by all remote entities, and data transmitted by the remote entities is only received by the central entity.

On the other hand the decentralised multi-end-point connection case is where data transmitted by any one entity is received by every other entity. In practice, this situation arises on typical broadcast networks. Important examples of this are certain satellite and local area network operations.

It should be noted that implementation of these cases is not trivial, because of the amount of synchronisation required to ensure 'simultaneous' establishment, disconnection, data transfer and acknowledge, etc. All the (N–1)-connections have to appear to the (N)-entity as a single (N–1)-connection, as far as the (N)-entity is concerned. The difficulty or otherwise of achieving this, depends on the underlying technology, i.e. whether it provides inherently a broadcast medium.

3.6   OTHER COMMON ELEMENTS OF LAYER OPERATION

In addition to the common architectural functions already discussed (like segmentation, blocking, concatenation, etc.) certain other functions may occur in several layers, and are thus best described in this general section.

3.6.1   Connection establishment

In the case where connection-mode services are being used at every layer, before Layer N can establish an (N)-connection, an (N–1)-connection must be available. If at the time of wishing to establish an (N)-connection, the (N–1)-connection is not, in fact available, it must first be established.

It follows therefore, that in the case where none of the layers have established connections, any request from an application must ripple down to the lowest layer and cause connections to be established from the lowest layer and upwards, one layer at a time.

Now there are cases in which the (N)-layer may find a connection already established at the lower Layer (N–1) in which case the (N)-connection may be established immediately (assuming such a connection is suitable). This situation can arise in two ways. Firstly because multiplexing is taking place and thus several (N)-connections are being 'threaded' down one (N–1)-connection. Secondly, it is sometimes more convenient to pre-empt higher layer requests and some layers have semi-permanent connections irrespective of whether a higher layer request has been received. This is particularly true at the Physical and Data Link Layers, where a permanent connection has no cost penalty and reduces overall connection establishment times, and reduces inter-layer management complexity.

Efficiency considerations have led to the concept of embedding, i.e. the carrying of (N+1)-connection requests as user data inside an (N)-connect request. In the extreme this would result in the simultaneous establishment of the seven layer connections. This, however, is not entirely practical for the following reasons:

(a) If embedding at all layers were permitted, then the size of the connect request for a given layer would need to take account of all the higher layers' requests. To take account of evolution, addition of new parameters, etc, this would make a requirement for arbitrarily long connect requests (especially at the lower layers).

(b) When multiplexing and/or re-use between Layer (N) and (N+1)-connections is taking place, embedding could only be used for the first (N+1)-connection. Thus, non-embedded (N+1)-connect requests would still have to be handled, producing two styles of operation and adding complexity with little benefit.

(c) Similarly, difficulties appear when Quality of Service (QOS) enhancement is required between Layer (N+1) and Layer (N), since the optimum (N+1) QOS function can only be applied when the knowledge of the resulting QOS has been obtained from the prior establishment of the (N)-connection.

However, there can be no doubt that seven independent layer connection establishments may be very time-consuming. Since the complications arise from multiplexing, re-use, or QOS it thus makes sense not to embed between layers where these functions occur, and to embed where these functions do not occur. This results in embedding being recommended for Layer 7 within 6, 6 within 5, with no embedding between 1, 2, 3, 4 or 5.

3.6.2 Connection release

Generally, an (N)-connection may be released by either of the (N+1)-entities, i.e. from either end, irrespective of which end originally initiated the connection.

Additionally there are circumstances in which the (N)-entities themselves can initiate release of the (N)-connection. This usually occurs as a result of some exceptional condition occurring in the (N)-Layer itself, which means that the (N)-connection cannot be maintained despite the fact that the (N+1)-entities still require it.

There are circumstances in which the release of an (N)-connection may result in loss of data still in transit (i.e. not yet delivered to the destination (N+1)-entity). This is because one type of release has precedence over all other possible actions. This has to be so, in order that malfunctions or malicious operation cannot prevent connections being released.

A second type of release termed an 'orderly' release in which the (N)-layer ensures that all data that has been passed to it is acknowledged by the remote entity before actually releasing the connection. In this case, the connection is released as far as the local (N+1)-Layer is concerned but the (N)-layer protocol connection remains until all data from the local (N+1) entity has been sent and acknowledged by the remote (N)-entity.

A third type of release is a negotiated release which extends the concept of an orderly release to include the remote peer (N+1)-entity rather than entrusting release to the (N)-Layer. Thus, unlike the orderly release where only data in one direction is guaranteed, a negotiated release requires agreement of both ends and thus can be used to guarantee data in both directions.

The effect that a release of one layer connection has over other layer connections depends on whether the following functions are being applied:

(a) multiplexing;

(b) splitting;

(c) orderly release;

(d) recovery mechanisms;

(e) cost optimisation.

In addition, it depends on the original connection establishment conditions (i.e. permanent or semi-permanent connection conditions described in

Section 3.6.1). Generally, local management decisions would be applied where specific functions do not dictate the inter-layer effects.

3.6.3 Transfer of data

The general relationships between service data units, protocol data units, etc. has already been described in section 3.3.3.

For connectionless-mode operation, once a connection has been established (N)-service data units may take arbitrary, but finite sizes (and we have seen how segmentation and reassembly are used to accommodate this).

There are other circumstances in which data in considerably smaller amounts may be transferred.

Some (N)-Layer services and protocols permit small amounts of (N)-user data to be transferred during (N)-connection establishment and/or release. The embedded connection establishment discussed in Section 3.6.1 is one use of such a capability. Typically, however, the amount of data is restricted, e.g. 32 to 128 octets. Other uses are for small amounts of management information (e.g. reasons for connection refusal, or for the network connection management subprotocol (described in section 5.3)) by the (N+1) layer.

For connectionless-mode operation the size of data-service units is not arbitrary. Since connectionless-mode operation is intended to provide a simpler service than connection-mode operation, the functions of the associated protocols are restricted in their capabilities. For example, segmentation and reassembly is only permitted in the Network Layer. Furthermore, in order to keep such segmentation/reassembly on a simple level, the maximum size of the network service data unit is 65 536 octets.

Above the Network Layer the service data unit sizes will decrease for each layer according to the amount of layer protocol control information.

3.6.4 Flow control

Flow control enables the amount of data being transferred to be regulated according to resources available in the end-systems or intermediate systems.

Between systems the flow control operates on the protocol data units. The mechanisms to provide flow control may vary according to the particular protocol design. Typically, flow control may be exercised by 'window' mechanisms as found in High Level Data Link Control (HDLC) and the packet level of CCITT Recommendation X.25, or by 'credit' mechanism as found in the Transport Layer protocols.

Clearly, the effects of, and requirement for, flow control apply between adjacent layers. However, the manner in which this is implemented is not standardised and conceptually operates on interface-data-units being transferred across the layer service boundary.

3.6.5 Expedited data transfer

The flow control procedures described in section 3.6.4 may under some circumstances result in an unacceptable cessation in the flow of normal data. Provision is made in some layers for an extra 'channel' of communication within the same (N)-connection in order to by-pass such a blockage, and consequently request what action to take in respect of the blockage.

To implement such a scheme it is clearly necessary to reserve resources to transmit and receive this Expedited data even where resources for handling normal data have been totally consumed. Thus, severe restrictions are made on the capability for Expedited data transfer to avoid imposing even more resource problems. The amount of data is restricted, 128 octets at the Network Layer, less at higher levels, and usually only one Expedited data unit may be outstanding at a given time. This service can only be viewed (and used) as an emergency channel. It cannot be used to provide two streams of data at different priority levels.

An Expedited data unit is a service data unit (i.e. no segmentation/ reassembly) which is transferred and/or processed with priority over normal service data units, and is independent of the status and operation of the normal data flow. The service provided by the Expedited data transfer guarantees that an Expedited data unit will not be delivered after any subsequent normal service data unit sent on the same (N)-connection. The intent is that in practice, the Expedited data unit will arrive before previous submitted normal service data units in the case where such units are being flow-controlled.

Since Expedited data transfer requires certain fixed relationships to be maintained with normal data transfer it is only applicable to connection-mode operation and *not* connectionless-mode operation (which because of its very nature cannot guarantee any particular order of delivery).

The use of Expedited data transfer among co-operating adjacent layers raises some interesting problems.

For example, the (N)-layer may require the (N−1)-Layer Expedited data transfer service for its own control purposes. Additionally, this (N)-layer may offer an (N)-Layer Expedited data transfer service to the (N+1)-Layer. Clearly, a conflict and contention for use of the (N−1)-Layer Expedited data transfer can now occur.

If, however, the (N)-layer offering an (N)-Expedited data transfer service to the (N+1)-layer does not use the (N–1)-layer Expedited data transfer service, then the (N)-Layer Expedited data can only be sent as (N–1)-Layer normal data. In this case, the Expedited effect can only occur within the queues dedicated to the particular (N)-layer and not within (N–1)-layer queues. Thus, blockage of (N–1)-Layer data also blocks (N)-layer Expedited.

If multiplexing of (N)-connections on to an (N–1)-connection is taking place, further complications arise, if mapping of (N)-Expedited data transfer on to (N–1)-Expedited data transfer is attempted. Lack of response to a single (N–1)-Expedited data transfer will preclude any other (N)-connection using the (N)-Expedited data transfer service.

These considerations have led to the following interlayer relationships, for Network, Transport and Session Layers:

(a) The Network Layer may or may not provide an Expedited data transfer service (i.e. it is not a mandatory service).

(b) Where the Network Layer does provide the Expedited data transfer service it can only be used by the Transport Layer if the Transport Layer is not performing multiplexing.

(c) The Transport Layer always provides an expedited data transfer service to the Session Layer. If the Network Layer Expedited is not being used (see (b) above) or not available (see (a) above) then the Transport Layer Expedited data transfer is transmitted via Network Layer normal data transfer. The Transport Layer itself does not require Expedited data transfer for its own control purposes. The Expedited data transfer service is offered to the Session Layer to primarily overcome any blockage occurring within the Transport Layer itself, or across the Transport/Session boundary.

(d) The Session Layer offers an Expedited data transfer service to the Presentation Layer, and in addition the Session Layer protocol requires use of the Transport Layer Expedited data transfer. Thus, Transport Layer Expedited data transfer is used for two purposes, one to provide the Session Service and two, for Session Layer control purposes.

(e) The Presentation Layer offers an Expedited data transfer service to the Application Layer.

It can be seen that the capability of an Expedited data transfer service throughout all the layers is not available and thus there is no guarantee that lower layer flow control mechanisms can be bypassed. In practice, this should not present a problem since the long-term blockages are more likely to occur at the application orientated layers and not within the generic communications oriented layers.

3.6.6 Sequencing

Sequencing is the term used to describe the property that data delivered by a given layer to its receiving user is in exactly the same order as it was submitted to the layer by the transmitting user.

This property is currently associated with connection-mode operation, and therefore use of a connection mode service automatically implies provision of sequencing. In most cases, the user of a connection-mode service would assume that sequencing was guaranteed. Only in extreme circumstances would the user of such a service (the next higher layer) take further measures to provide sequence integrity mechanisms.

Connection-mode services, on the other hand, do not give any guarantee of sequenced delivery. If, therefore, an (N)-Layer connection-mode service is to be provided over a (N+1)-Layer connectionless-mode service, appropriate sequencing mechanisms must be included in the (N)-Layer. Normally, the (N)-protocol-control-information would contain appropriate sequence numbers for every (N)-PDU and there would be procedures for detecting loss, duplication, and subsequent recovery, to provide an eventual 'in-sequence' delivery. Examples of such mechanisms may be found in the Transport Layer (class 4 protocol for use over the connectionless-mode network service) and in the Data Link Layer (HDLC procedures to deal with losses due to transmission errors).

It should be noted that an unsequenced connection-mode service could be provided, and, indeed, has been raised for certain applications. Were this to be provided, sequencing could be regarded as a QOS parameter of the connection-mode service. More will be said on this topic in Chapter 4.

3.6.7 Reset

Reset is a layer service provided by the Network and Data Link Layers to interrupt the operations of a layer, destroying any data held within the layer (i.e. pending delivery to the service user or the lower layer), and resetting the state of the connection (internal counters, timers, etc.) to a predefined state. Reset is needed to overcome drastic but temporary failures in the communications path (such as loss of a relay) and to inform the service users that this has occurred. Resets can be initiated by service users or service providers.

3.6.7.1 Service-user initiated reset

The probability of a flow control blockage of normal data transfer, and the use of Expedited data transfer as a means of effecting removal of the blockage, was considered in section 3.6.5.

The use of Expedited data transfer may not always be appropriate nor, indeed, guarantee removal of such a blockage. It will not be appropriate for removing blockages from intermediate systems, for example, which do not interpret end-to-end Expedited-data transfers. It may also be the case that end systems do not employ Expedited or it has been already tried without effect. In these cases, a more drastic action is required to remove the blockage.

If the blockage cannot be removed by the receiver processing the queue of accumulated data, clearly the only remaining way to remove the blockage is by destroying the data in the queue. Thus, a user-generated reset has the property of destroying all data between the user generating the reset and the remote peer user receiving the reset.

3.6.7.2 Service-provider generated reset

In addition to the deliberate destruction of data incurred by the use of the Service-user generated reset above in section 3.6.7.1, there may be certain abnormal conditions within the layer itself which lead to loss (or possibly duplication) of data accidentally (e.g. protocol malfunctions, buffer/queue management problems). In the circumstances, the entities within the layer need to be able to 'restart' their respective protocol machines to pre-defined states if meaningful communication is to continue. In this case both peer users are also informed (by Reset Indications). This enables the users to be aware that data loss (or duplication) may have taken place, and taken appropriate higher-level recovery action if required.

3.7 RELAYING

The reference model defines an (N)-relay as:

> 'An (N)-function by means of which an (N)-entity forwards data received from one correspondent (N)-entity to another (N)-entity.'

Thus, generally speaking, in the case where A is communicating to C via B where A, B, C are all entities within the same layer, B is said to be performing relaying. A simple requirement for this would be in the case where A and C have different protocols and B is acting as a protocol converter, as shown in fig. 3.17.

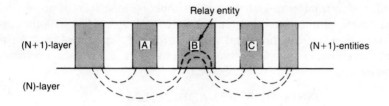

Fig. 3.17 Relaying in the (N)-layer.

There are instances where protocol conversion cannot be performed directly within a given layer, or the service of a given layer does not permit end-to-end communication. However, (N+1)-entities can only intercommunicate by using the service of the underlying (N)-layer, which in these instances contains a discontinuity. In this case, communication between the end-systems can still be effected if some other intermediate (N+1)-entity can act as a relay. This can is shown in Figure 3.18.

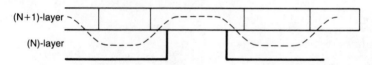

Fig. 3.18 Discontinuity in the (N)-layer.

The Reference Model, however, does place restrictions on the layers within which relaying may take place. The prime reason for this arises from the need to determine the precise semantics of confirmed services. That is, where a protocol provides a response of some kind, e.g. an (N)-Connect confirm, or a piece of protocol acknowledgement mechanism, the receiver must not be left in any doubt as to the origin of the response. The Transport Layer mechanisms are meant to provide end-to-end assurance for a number of services. Since the insertion of an intermediary would destroy the end-to-end assurance, relaying within the Transport Layer is not permitted. It also follows that no relaying can occur above the Transport Layer for a single instance of communication (there can be cases of relaying by the Application Layer, see below). In the case of two end-systems communicating only with each other for a given application, relaying will only take place below the Transport Layer.

There are situations which involve more than a single pair of end-systems, e.g. distributed applications, electronic mail applications, etc. For these applications, relaying at the level of the Application Layer will be necessary and this is permitted. Such relaying may be of a real-time nature or of a store-and-forward nature. The latter case should not be confused with connectionless-mode operation, simply because there is no real-time end-to-end connection between the source and destination in some cases. Connections may be used and will exist for the duration of the transfers and interactions between the store-and-forward Application entities themselves.

3.8 THE CONNECTIONLESS ADDENDUM

The Basic Reference Model ISO 7498, described thus far, is primarily concerned with connection-orientated operation. An Addendum has been produced to cover connectionless-mode transmission. This Addendum has been designated ISO 7498/AD1.

The best way to explain connectionless-mode operation initially is by contrasting it with connection-mode operation.

For connection-mode operation an association is established between peer-entities and between each peer entity and the next lower layer (refer back to Fig. 3.5). Connection-mode operation involves the following three distinct separate phases:

(a) connection establishment;

(b) data transfer;

(c) connection release.

The following fundamental characteristics are exhibited by connection-mode operation:

(a) The establishment of a three-party agreement between the peer-entities and the underlying layer for all related parameters which are to be applied during the data transfer phase.

(b) Means of connection identification, obviating the need for overheads for data transfer.

(c) There is a context within which individual units of data passing between the peer-entities can be logically related. This makes it possible to provide sequencing and flow control mechanisms.

(d) The context described in (c) has a distinct beginning and end, coincident with connection establishment and release.

Connectionless-mode transmission does not include any distinct phases. There is no connection to be established, no data phase, and nothing to be released. Each unit of data (termed (N)-Unitdata) is entirely self-contained, and all the information required to deliver a unit of data (destination address, QOS selection, options etc.) is presented to the layer providing the connectionless-mode service, together with the data, in a single service primitive. The layer providing the connectionless-mode service is not required to relate any one service access (by such a primitive) to any other.

The same basic architectural principles apply to connectionless-mode as were described for connection-orientated operation as described in sections 3.1 to 3.4. Many of these are repeated in the connectionless-mode addendum but will not be repeated here (e.g. service access points, service data units, interface units, protocol data units, relays, etc.), since they have already been covered.

As far as QOS is concerned, two basic categories of QOS parameters are defined: dynamic and static. Dynamic QOS parameters are those that can be set for a particular invocation of the service and include items like priority, transit delay, etc. Static parameters apply for multiple invocations of the service (long-term) between a pair (or pairs) of (N)-service access points. Characteristics like sequence preservation probability, duplication probability, etc. fall into this category. These characteristics are made known to the service user by local arrangements. Such local arrangements can be modelled by the provision of a management channel between the layer providing the service and the layer using the service. This management channel permits the service user to enquire about the characteristics using an (N)-facility request primitive, to which the provide replies with an (N)-facility indication as shown in Fig. 3.19. Indications may be initiated independently of the requests under some circumstances, layer initialisation, for example.

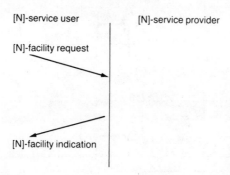

Fig. 3.19 Inter-layer management channel.

The main new aspects dealt with in the Addendum relate to the multi-layer and inter-layer relationships between connectionless-mode and connection-mode operation.

For a given layer, the service offered to the two communicating end-systems has to be the same. Thus, interworking between services of different types requires that one of the services (either the connection-mode or the connectionless-mode one, depending upon economic and technical factors) be first converted to the other.

The basic cases of relaying are illustrated in Fig. 3.20. In order to maximise the possibility of interworking and to limit protocol complexity, a restriction has been placed on the number of layers within which conversion between one mode of service and the other may take place. Firstly, no conversions may take place above the Transport Layer. Secondly, conversion may take place in the Transport Layer itself, under specific conditions, as follows:

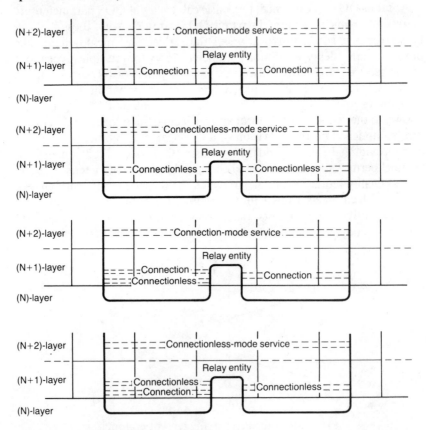

Fig. 3.20 Connectionless/connection-mode relaying; (a) relaying between two connection-mode services; (b) relaying between two connectionless-mode services; (c) connection- and connectionless-mode services yielding a connection-mode service; (d) connection- and connectionless-mode services yielding a connectionless-mode service.

A system which is intending to be fully open must support a given mode of Transport service over a Network service of the same type (utilising conversion within the Network Layer if necessary). Such a system may, in addition, provide conversion in the Transport Layer.

That is, a system which supports the connection-orientated Transport Service must support the connection-orientated Network service. Similarly, a system which supports the connectionless-mode Transport Service must support the connectionless-mode Network service.

Despite these limitations, imposed to try to minimise incompatibility between implementations, a number of manufacturers have chosen to provide the crossover in the Transport Layer, providing the connection-orientated transport service over the connectionless network service.

Furthermore, in order to limit protocol complexity (to indeed make the connectionless-mode operation simple and not duplicate the functionality of connection-mode operation), segmentation and reassembly is not permitted above the Network Layer. In consequence of this, the size of service data units in layers above the Network Layer is limited by the size of service data unit provided by the service of the layer below and by the size of the protocol control information for the layer itself.

Special considerations apply to the Physical and Data Link Layers. Connection-mode and connectionless-mode services are not differentiated for the Physical Layer. The services of the Physical Layer are determined by

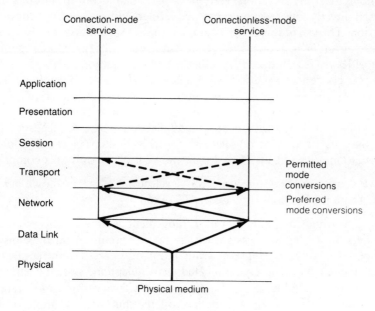

Fig. 3.21 Preferred and permitted mode conversions.

the characteristics of the underlying medium and are too diverse to allow categorisation into connection-mode and connectionless-mode operation. Functions in the Data Link Layer must convert between the services offered by the Physical Layer and the type of Data Link Service needed, which may be either mode.

The permissible conversions are shown in Fig. 3.21.

3.9 THE NAMING AND ADDRESSING ADDENDUM

General naming and addressing principles are specified in a further Addendum to IS 7498: (Part 3).

The Addendum contains a set of formal definitions which are vital to the understanding of the concepts, and correct use of terms.

A name denotes (identifies) an entity to which it is bound. Names themselves are classified into three types for particular usage within OSI:

(a) a *title*, which names a layer entity or an application process;

(b) an *address*, which names a group of service access points all located within a single end-system;

(c) an *identifier*, which names some other kind of object.

It follows that only an address identifies a particular location. A title cannot be used directly, but must be translated into an address by some directory function. The use of directories is limited to the Application and Network Layers, in order to reduce complexity and thus only two global naming authorities are required, one for allocation of Application titles and one for network service access point (NSAP) addresses. The scope of a middle layer address is restricted to be valid only within the scope of the underlying SAP. This produces a simple address fan-out arrangement as shown in Fig. 3.22. The limited use of the Transport and Session addresses leads to the introduction of the term 'selector' for this limited piece of address information. The address of an Application entity can only ever comprise:

NSAP address + TSAP selector + SSAP selector + PSAP as shown in Fig. 3.22. There may of course be many NSAPs in a given system.

A very rough analogy to this arrangement is the use of extension numbers for a telephone system within a user's premises. Such extension numbers are not globally useful or unambiguous, except in the context of the global telephone number allocated to the telephone system itself.

The number of addresses and selectors used will vary from end-system to end-system and will be a function of the number of applications, application invocations and internal structure of the end-system itself.

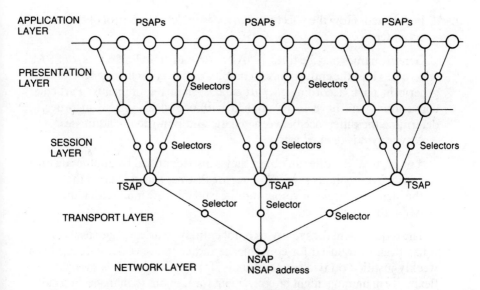

Fig. 3.22 OSI multilayer address components.

Many end systems will be highly complex and may comprise many interconnected pieces of equipment, each containing complex software structures. Selectors are needed to locate items of software within such end-systems. Selectors are needed in Layers 4 to 6 to allow for the different distribution of software elements possible in each end-system.

Often, the end-system will be rather simple and only need selectors between the software provided by the manufacturer (e.g. operating system software) and the many user-written applications. The location where a selector is used will depend on how much of OSI has been provided by the manufacturer, and in particular the highest layer provided by the manufacturer.

The Application Layer directory function provides a mapping from an Application title into the PSAP address required to access the Application-entity identified by that title, i.e. an NSAP address, Transport selector, Session selector and Presentation selector. The use of selectors is not mandatory and direct mappings between titles and TSAP addresses, or NSAP addresses, is permitted.

To cover distributed systems, resource sharing systems, or systems replicated for reliability purposes, provision is made for use of generic names. A generic name is:

the name of a set of objects such that, when used to specify the recipient of a communication, its effect is that exactly one member of the set will be selected. How the selection is made is under the control of the service provider.

Generic names may be titles, NSAP addresses, TSAP selectors, or SSAP selectors. Thus, the entire address combination may itself be generic or only generic in part, according to particular system requirements. Particular conventions will have to be applied within a given system in order to determine the difference between generic and non-generic addresses.

A synonym is defined as:

a name that identifies an object that is also identified by another distinct name. Synonymous generic names are distinct generic names that name the identical set; synonymous multi-cast names are multi-cast names that name the identical set.

The requirement for synonyms is not entirely clear, and one suspects that it has been introduced for convenience rather than necessity. It has been weakly justified on two counts. Firstly, for applications it is thought that flexibility in naming might be convenient for humans to manage. Secondly, in order to avoid complex routing problems, the allocation of several names for the same physical end-system could be used to distinguish different subnetwork paths to that end-system. Synonyms can also be applied to generic names as per the definition above.

The directory function for the Network Layer provides for the mapping between an NSAP address and the routing information required below the Network Layer service boundary in order to create a path to the appropriate destination NSAP. A more detailed treatment of Network layer addressing will be given in Chapter 4.

As far as layer protocols are concerned, the address information for a given layer is always conveyed within the protocol of that layer.

PART 2

INTERCONNECTION OF OPEN SYSTEMS

CHAPTER 4

Layers 1 to 3

4.1 GENERAL REQUIREMENTS OF THE NETWORK LAYER SERVICE

In an ever-changing technological environment it is vital that there is a practical and flexible means of interconnecting all possible networks. The higher layers must be insulated as far as possible from the variety of technologies that may make up a composite path across a number of networks, and this principle generates two criteria that the general interworking mechanisms must satisfy.

First, higher-layer functions in the communicating end-systems must have a common understanding of the nature of the service available between them so as to construct higher-layer protocols upon a stable basis. Clearly, this basis must be independent of the types of communications media between the end-systems, and also independent of whether or not composite communications paths across a number of interconnected data networks are employed. In both ISO and CCITT there is agreement that it is the task of the Network Layer of the OSI model to provide this common service, which is therefore called the OSI Network service. Gateway functions at higher layers must not be needed to accommodate new communications technologies, or the general-purpose procedures developed at the Transport Layer and above will find themselves restricted to particular types of communications media, which would be a quite unworkable prospect.

It is worth re-emphasising here that the term 'Network service' has to be understood as the service provided to Transport Layer functions in end-systems. It is emphatically *not* the service provided by any particular network. Furthermore, service has to be understood as meaning the functions available between end systems – e.g. connection establishment, data transfer, reset mechanisms – not in terms of quality of service (QOS) such as residual error-rate.

The second criterion that the interworking methods must satisfy is that end-systems should not need to modify their procedures to communicate over composite paths. Of course different addresses will apply, but the protocols operated by a particular end-system should be the same whether that end-system is communicating 'locally' across its own network or is using a composite path half way round the world.

Interworking several different networks together to form a single super network is not simply a matter of connecting them together. To avoid confusion, real networks will be individually termed 'subnetworks', to distinguish them from the super network required for OSI.

4.2 SUBNETWORK INTERCONNECTION

Consider the configuration in Fig. 4.1. Systems A and B communicate with each other using the facilities offered by subnetwork (in our terms) X; and systems C and D communicate using the facilities of subnetwork Y. The higher-layer functions in systems A and B rely on the availability of the

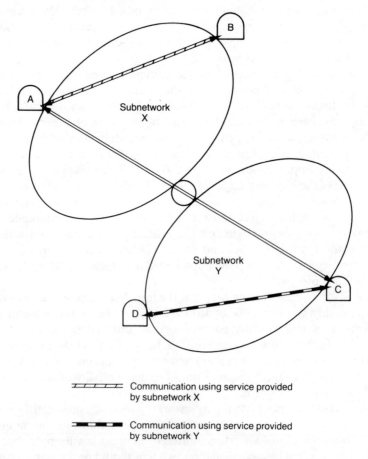

Fig. 4.1 The interworking problem.

facilities of subnetwork X; those in C and D rely on the availability of the facilities of subnetwork Y. Now suppose that A needs to communicate with C. There is, in general, a mismatch between the facilities provided by the two subnetworks. This mismatch is typically not just a difference of encodings, which can be resolved easily, but a fundamental difference in what is provided. It is then not possible to build a gateway between X and Y that converts X-facilities to Y-facilities, on a one-to-one basis.

Typically, the gateways that have been built in practice convert between different encodings of the facilities that are common to the two networks concerned, and ignore or preclude the others. The immediate consequence of that is that the facilities available between A and C are a subset of those available between A and B; and so the higher-layer functions in A have to distinguish between local communication and interworking. A less immediate but very serious drawback is that if it is required to communicate from A to a system on some other subnetwork Z, in general the set of facilities common to X and Y in combination will be different from those common to X and Z. Then the higher-layer functions in system A have to distinguish between local communications and two different cases of interworking.

Clearly, the above approach does not give a solution to the general interconnection problem. The general interconnection problem demands a general solution.

OSI provides a general solution. The OSI Network service is a definition, in abstract terms, of what higher layers may expect from the communications layers: it specifies a universal common set of facilities that interworking methods should support. Subnetwork services differ, and interconnection of subnetworks is difficult; but when enhanced to the OSI Network service, the enhanced subnetwork services are the same, and interconnection of enhanced subnetworks is straightforward.

In architectural terms, the recognition of this difference has led to the concept of the OSI Network service and the subnetwork over which it is being provided, as shown in Fig. 4.2. This concept has led to the development by OSI of a separate architecture for the Network Layer, which is contained in ISO 8648 – Internal Organisation of the Network Layer.

4.3   INTERNAL ORGANISATION OF THE NETWORK LAYER (IONL)

An architecture is necessary because of the varying capability of real network technologies (termed subnetwork in OSI) to fulfil the OSI Network service, and different interconnection strategies.

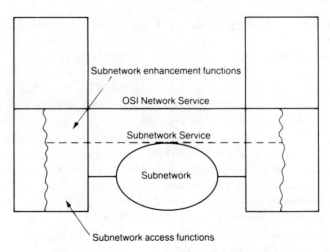

Fig. 4.2 Subnetwork enhancement.

Many existing real network technologies require an over-the-top protocol element if they are to be used to provide the OSI Network service. Additionally, gateways between subnetworks can be constructed to operate on different levels of Network Layer protocol, and two distinct methods of operation have been identified. These issues give rise to a need to be able to speak formally about a given protocol structure, and to specify the purpose for which a given protocol is being used. This architecture, in effect, causes the Network Layer to be subdivided in the sublayers mentioned previously in section 3.1.

To achieve the architectural requirements the notion of 'roles' is introduced.

Three roles for Network Layer protocols have been identified to describe how multiple Network Layer protocols may be employed to construct the OSI Network service:

(a) in the role of subnetwork-independent convergence protocol (SNICP);

(b) in the role of a subnetwork-dependent convergence protocol (SNDCP);

(c) in the role of a Subnetwork access protocol (SNAcP).

A protocol within the Network Layer may fulfil one of these roles in a particular configuration; the same protocol may fulfil the same or different roles in different configurations.

It is not necessarily the case that each role is fulfilled individually by a separate discrete protocol; in particular, a single Network Layer protocol may provide all of the functions that would have been provided by protocols operating in the other roles.

4.3.1 Subnetwork-independent convergence protocols (SNICP)

The SNICP role exists where the construction of the OSI Network service is achieved by a protocol based on a minimum level of service from a subnetwork.

Such a protocol is designed to be usable over as many subnetworks as possible, thus the assumptions it makes about underlying services are minimal. In fact, an artificially primitive service is defined for the underlying service, which all known subnetworks are known to meet with ease.

Any useful features over and above this minimal service, actually available from a subnetwork, will thus be ignored when a SNICP is in use.

4.3.2 Subnetwork-dependent convergence protocols (SNDCP)

A protocol fulfilling the SNDCP role operates over the SNAcP. It will either provide the OSI Network service or exceptionally provide the service required for the SNICP.

It follows from the description of SNICP above that since in general the service required by a SNICP is always satisfied by any subnetwork, SNDCP is not required if SNICP is being used. Only exceptionally will an SNDCP be required with an SNICP.

Normally, it is expected that an SNDCP will be used to realise the OSI Network service directly.

4.3.3 Subnetwork access protocols (SNAcP)

This is the native protocol of the subnetwork itself, and operates in accordance with the conditions stated explicitly for the characteristics of a specific subnetwork.

Since the level of service that such a protocol contributes towards the OSI Network service is specific to the subnetwork concerned, the subnetwork may or may not coincide with the OSI Network service.

Readers familiar with the CCITT Recommendation X.25 will realise that the 1980 version of the recommendation is a SNAcP which does not fulfil the OSI Network service, whilst the 1984 version does.

In general, the way in which a SNAcP contributes to the realisation of the OSI Network service is governed by the level of service provided by the specific subnetwork and additional Network Layer protocols, if any, required to make-up the OSI Network service.

4.3.4 Relaying and routing

Relaying functions enable a Network Layer entity to forward information received from one correspondent Network Layer entity to another correspondent Network Layer entity, i.e. those functions resident in intermediate systems, without which the other Network Layer entities on either side could not directly communicate.

This situation can be illustrated as shown in Fig. 4.3. The 'string' of Network Layer entities require relays to communicate with one another because the Data Link Layer only communicates between pairs of adjacent entities. The figure is drawn to capture the physical aspects as well as the normal abstract architectural ones, i.e. the disjoint nature of the Data Link Layer in a multi-node configuration.

Routing functions in intermediate systems determine the path over which the data is to travel. These functions are concerned with the interpretation of addresses, quality of service parameters, local state information, and/or

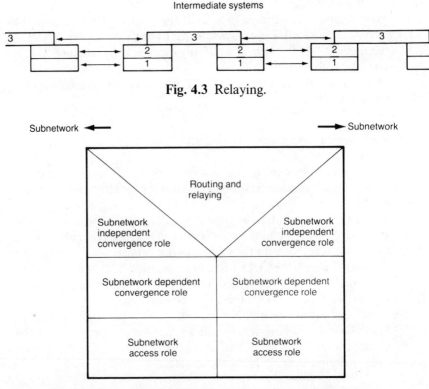

Fig. 4.3 Relaying.

Fig. 4.4 Generalised architecture for an intermediate system.

other parameters. Routing is a complex subject, and many different routing algorithms and strategies are possible. Basically, however, the process is to determine which neighbour to choose out of a set of neighbours, for the required communication path for a given destination address.

These concepts concerning roles, protocols fulfilling the roles, and relaying and routing lead to the generalised Network Layer architecture shown in Fig. 4.4. In the figure, intermediate systems are seen as comprising two halves, each half looking towards a particular subnetwork consisting of the subnetwork access functions and enhancement functions particular to that subnetwork. Putting the two halves together needs additional functions to map between the different encodings used in the two halves, and also routing and parameter management functions needed to cater for the multiplicity of paths.

It should be noted that when multiple real subnetworks are present in a given configuration there are two approaches to the representation of the relaying and routing functions in individual Network Layer entities, depending on the degree of abstraction required. The various cases are shown in Fig. 4.5(a) to (c). A summary of related terms is shown in Fig. 4.6.

4.3.5 Subnetwork interconnection strategies

We can now return to the strategies for interworking, i.e. provision of the Network service over an interconnection of different real networks. Recalling the discussions on SNICPs and SNDCPs there are two basic strategies for the following protocol combinations:

(a) SNAcP + SNDCP, known as the 'hop-by-hop enhancement' approach;

(b) SNAcP + SNICP, known as the 'internet' approach.

4.3.5.1 Hop-by-hop enhancement approach

The approach hop-by-hop enhancement takes each subnetwork in the chain individually and, as necessary, enhances its subnetwork service by the addition of Network Layer functions, up to the level of the OSI Network service.

This requires the operation of, over each interconnected subnetwork, an SNDCP enhancement protocol tailored to that particular type of subnetwork, and then provide the necessary mappings between different but equivalent implementations of the OSI Network service in 'gateways' or 'relay units' between the subnetworks. This is illustrated in Fig. 4.7.

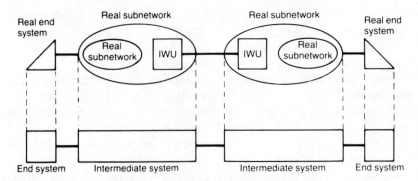

Fig. 4.5(a) Representation of each real subnetwork and its interworking function as an intermediate system.

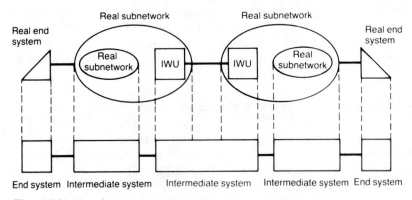

Fig. 4.5(b) Representation of combined interworking functions as a separate intermediate system.

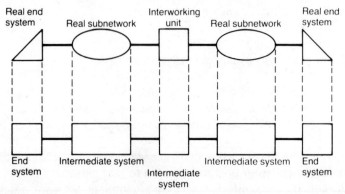

Fig. 4.5(c) Interconnection real subnetworks by a separate interworking unit.

Real world object	Graphical representation of real world object	Corresponding abstract element	Graphical representation of abstract element
Real system		System	
Real open system		Open system	
Interworking unit	IWU	Relay system	Intermediate system
Real subnetwork	Real subnetwork	Subnetwork	Intermediate system
Real end system	◿	End system	End system

Note: The abstract term 'intermediate system' is used to refer to either a network layer relay system or a subnetwork.

Fig. 4.6 Summary of related terms.

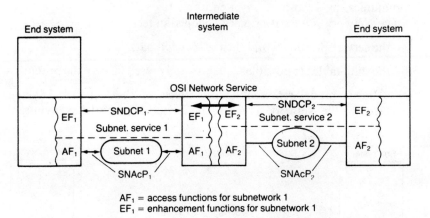

AF₁ = access functions for subnetwork 1
EF₁ = enhancement functions for subnetwork 1

Fig. 4.7 Hop by hop enhancement.

There is a special and important degenerate case of this strategy which occurs when one or more of the subnetworks is capable of conveying all the primitives and parameters of the Network service. No enhancement protocol mechanisms are needed over such subnetworks. (The more recent 1984 revision to CCITT Recommendation X.25 is an example of this.)

4.3.5.2 Internet approach

This strategy involves the operation of the same enhancement protocol (i.e. a single SNICP) called an 'internet' protocol, across all the interconnected subnetworks. Such an internet protocol must itself support every feature of the Network service that will not be directly supported by every subnetwork over which it has to operate. It can therefore be thought of as an enhancement protocol tailored to the absolute minimum subnetwork service that might be encountered. This strategy is illustrated in Fig. 4.8.

The advantage of the hop-by-hop enhancement approach is that maximum use is made of the facilities of particular subnetworks, thus tending towards efficiency of operation. The advantage of the internet approach is that only one protocol need be implemented to cater for all circumstances, but at the expense of ignoring and possibly duplicating the facilities present in particular subnetworks.

4.4   OSI NETWORK SERVICE AND ITS PROVISION

We are now in a position to consider the characteristics of the OSI Network service, i.e. the service of the Network Layer available between the communicating Transport Layer entities.

The Network service is primarily defined in terms of:

(a) the service primitives and their associated parameters;

(b) the state-tables of possible sequences and precedences of the primitives;

(c) a queue model representing the effects of the Network Layer as perceived by the Network service users (i.e. the Transport Layer entities).

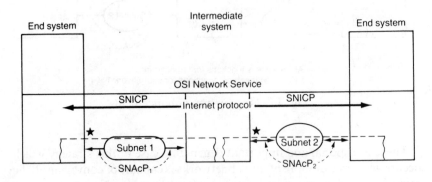

* = minimum subnetwork service

Fig. 4.8 Internet approach strategy.

4.4.1 Connection-mode Network service (CONS)

The CONS has been jointly developed by ISO and CCITT as ISO 8348 and Recommendation X.213, respectively. There are some minor differences between these two Standards, but the differences are such that interworking one with the other is possible, with only some slight loss of diagnostic information.

The Network service defines:

a connection which may be established or terminated between the Network service users for the purpose of exchanging data;

for each connection, when it is established, certain measures of quality which are agreed between the network service provider and the Network service users;

the transfer of Network service data units (NSDUs) on a connection; the transfer is transparent, in that the boundaries of NSDUs and the contents of NSDUs are preserved unchanged by the service, and that there are no constraints on the data values imposed by the service; the transmission of this data is subject to flow control;

the transmission of separate Expedited data, which is subject to a different flow control from the normal data;

means by which the connection can be returned to a defined state, and the activities of the two users synchronised, by use of a Reset service;

means for the service user to confirm receipt of data;

the unconditional and therefore possibly destructive termination of a network connection.

4.4.1.1 Connection establishment

Four primitives are used to establish Network connections, as follows:
(a) the calling Network service user (i.e. Transport entity) issues an (N)-Connect request;
(b) the called Network service user receives an (N)-Connect indication;
(c) the called Network service user may accept with an (N)-Connect response which appears at the calling Network service user as an (N)-Connect confirmation.

The called Network service user may choose to refuse the connection with an (N)-Disconnect request. The Network service provider may itself refuse with an (N)-Disconnect confirmation.

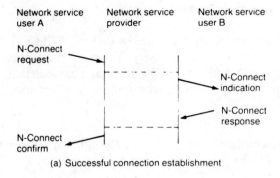

(a) Successful connection establishment

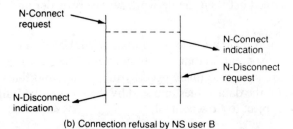

(b) Connection refusal by NS user B

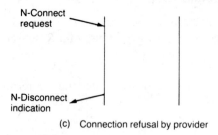

(c) Connection refusal by provider

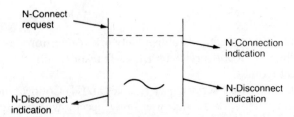

(d) Connection refusal by provider
subsequent to delivery of
N-Connect indication

Fig. 4.9 Connection establishment.

Parameter/ ⟋Primitive	(N)-Connect request	(N)-Connect indication	(N)-Connect response	(N)-Connect confirm
Called address	X	X(=)		
Calling address	X	X(=)		
Responding address			X	X(=)
Receipt confirmation selection	X	X	X	X(=)
Expedited data selection	X	X	X	X(=)
Quality of Service parameter set	X	X	X	X(=)
Network Service user data	X	X(=)	X	X(=)

Fig. 4.10 Summary of connection establishment primitives and associated parameters.

Figure 4.9(a) to (d) show the four possible cases.

Parameters associated with the establishment are:

(a) called and calling addresses, up to 40 digits each;
(b) Network service user data (for each direction) up to 128 octets;
(c) Quality of Service information;
(d) option selection.

A summary of the connection establishment primitives and their associated parameters is shown in Fig. 4.10. Where there is an (=) against the parameter X this indicates that the value of X is identical to that in the originating primitive (i.e. from request to indication, from response to confirm).

It should be noted that the address parameter in the response/confirm is termed the responding address rather than the called address. This reflects the possibility that under circumstances where redirection or generic addressing is involved, the actual destination responding to the connection request may be different from that originally addressed.

4.4.1.2 Quality of Service (QOS)

QOS parameters can be divided into three classes:

Class (a) – those which are negotiated between the two Network service users and the Network service provider (three-party negotiation);

Class (b) – those which are selected by the calling Network service user and met by the Network service provider and signalled to the called Network service user;

Class (c) – those which are not negotiated externally between end or intermediate systems (and thus have no protocol requirements) and whose values are selected and/or known by other means.

For class (a) values of QOS parameters requested by the calling Network service user may be modified in the course of call establishment in the forward direction by the Network service provider and/or the called Network service user. No modification is permitted in the reverse direction, however, thus values in the (N)-Connect confirm are identical to those in the (N)-Connect response. This restriction is necessary to ensure that both Network service users have the same understanding and agreement on the QOS.

For class (b), the values of QOS parameters may not be modified by the provider and consequently the values in (N)-Connect request and (N)-Connect indication are identical. If the Network service provider cannot meet the request, the call must be rejected.

For each QOS parameter subject to the three-party negotiation (class (a)), four further subparameters are defined. The calling Network service user is allowed to specify in the (N)-Connect request a 'target' value and a 'lowest acceptable' value. If the Network service provider cannot meet the lowest acceptable value then the call is rejected. The (N)-Connect indication specifies the 'available' value (possibly lower than the original target) and the lowest acceptable to the called Network service user. The called Network service user can reply with any value between the lowest acceptable *and* the available. The resulting reply is defined as the 'selected' value and is conveyed in the (N)-Connect response and confirm primitives.

The following parameters have been defined for each class:

| Class (a) | Throughput |
| (Negotiated) | Transit delay |

Class (b)	Priority
(Selected and	Protection
indicated)	

Class (c)	Connection establishment delay
	NC establishment failure probability
(Not negotiated	Residual error rate
nor selected/	Transfer failure probability
indicated)	NC resilience
	NC release duty
	NC release failure probability
	Maximum acceptable cost

4.4.1.3 Option selection

There are two optional features specified for the network service, namely Expedited data and receipt confirmation. These services have to be requested by the calling Network service user if they are to be used, and then subsequently agreed by the Network service provider and the called Network service user. The Network service provider is not obliged to implement these optional services but it is obliged to participate correctly in the negotiation of the features.

Four situations can arise for such Network service provider options:

(a) the calling Network service user does not request the option. In this case it is not available.

(b) the calling Network service user requests the option but the Network service provider (via local, all intermediate or remote Network entities) has not implemented the option. In this case the (N)-Connect request would have the option value set to 'use', but the (N)-Connect indication would have the option value set to 'no use'.

(c) the calling Network service user requests the option and it is implemented by the Network service provider (via local, all intermediate, and remote Network entities) but the called Network service user does not wish to use it. In this case both the (N)-Connect request and indication would have the option value set to 'use', but the (N)-Connect response has the value set to 'no use'. The (N)-Connect confirm value would also, therefore, have the value 'no use'.

(d) the calling Network service user requests the option, the Network service provider supports the option and the called Network service user agrees to its use. In this case the value in all primitives, including the (N)-Connect confirm, have the value set to 'use'.

4.4.1.4 Connection release

An established, or partially established, connection may be released by either Network service user, or by the Network service provider. Fig. 4.11(a) to (d) shows the possible sequences.

It should be noted that the user initiating the release with an (N)-Disconnect request does not receive any further indication of the release, i.e. the release is assumed to be instantaneous and its confirmation is implicit. Local implementations, may, of course, include local confirmations if required. The times at which a given implementation permits new (N)-Connect requests subsequent to (N)-Disconnect requests/indications is a local matter.

When a Network service user receives an (N)-Disconnect indication it contains originator and reason parameters. This enables the Network service user to know whether it was the remote Network service user who originated the release or whether it was abnormally terminated by the Network service provider (and, in the latter case, why).

Note: the CCITT Network service definition X.213 allows the originator parameter to take a value of undefined. Thus, in practice, implementations are well advised to take account of this.

In ISO 8348 the (N)-Disconnect request (and consequently the (N)-Disconnect indication) allows Network service user data (0 to 128 octets) to be transferred with the release of the connection. ISO 8348 Network service has this as a mandatory feature for the Network service provider. CCITT

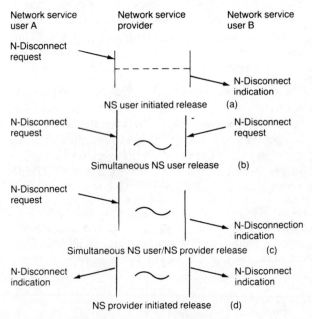

Fig. 4.11 Connection release.

X.213, on the other hand, has not made this mandatory, but rather as a provider-option currently. Data associated with release cannot be protected against loss because a remote part of the connection may have already been released, arising from simultaneous remote Network service user/provider generated release.

Furthermore, since there is no confirm for the release at the service level, the sending Network service user has no indication that it has been received by the remote Network service user.

It should be noted that any data still in transit but undelivered at the time a release is initiated may be destroyed. The same will be true for all other service elements (i.e. Expedited, reset, or data acknowledgements described later in this section).

For these reasons, and the CCITT optionality issue, the wisdom of employing this feature is questionable.

4.4.1.5 Data transfer

On an established Network connection, both Network service users can request simultaneously the transfer of user data via the user-data parameter of (N)-Data requests, which emerge at the other end of the connection as (N)-Data indications. The Network service data units (NSDUs) are delivered in sequence.

In the Network service definition, the NSDU length is unlimited (but must be an integral number of octets). The protocol supporting the service will, of course, contain the necessary segmentation/reassembly mechanisms.

Flow control is a property of the Network service but the way in which it is to be achieved is not specified. All that is said about flow control in the Network service, is that the Network service offers a means by which the receiving Network service user may control the rate at which the sending Network service user may send NSDUs. In practice, various protocols have particular methods of achieving flow control as will be seen later. These may be mapped directly on to layer interface flow controls, or may only have 'loose' relationships with local layer interface flow control mechanisms, at the designer's choice. Therefore, one can only abstractly describe flow control in terms of the 'back-pressure' effect.

As discussed in section 3.1, the Transport Layer is responsible for end system to end system assurance of data. Thus, in general, there is no requirement for the Network service to provide any form of data assurance. However, when X.25 was considered in relation to the Network service, there was a view that the use of the D-bit (X.25 section 4.3.3) to obtain an end-to-end acknowledgement of delivery (via the packet receive sequence number P(R)), should be reflected. This has resulted in the inclusion within the Network service, of receipt confirmation, as a provider option (i.e. only likely to be available if X.25 is in sole use). The Transport Layer, therefore, cannot rely on such a service always being available, and different classes of Transport protocol have been defined to cater for its presence or absence.

Receipt confirmation is requested by setting the Confirmation request parameter on the (N)-Data primitive. The confirmation of receipt occurs via the (N)-data acknowledge request/indication primitive in the reverse direction. The events are summarised in Fig. 4.12.

4.4.1.6 Expedited data transfer

The Expedited data transfer is a provider option which may not always be available. Where it is available its usage has to be agreed between the two service users, i.e. the Transport entities, during the connection establishment procedure. Either the provider or the remote user may refuse the request for Expedited.

This option reflects a difference between packet and circuit switched network characteristics. Expedited is deemed to be necessary in packet switched networks because of the greater potential for blockage of data. Conversely, it is deemed unnecessary in circuit switched networks which provide transparent bit pipes end-to-end and thus cannot create blockages.

The amount of Expedited data that can be transferred is limited to a maximum of 32 octets. Any submitted Expedited data, remaining undelivered at the time a release or reset is initiated, may be destroyed.

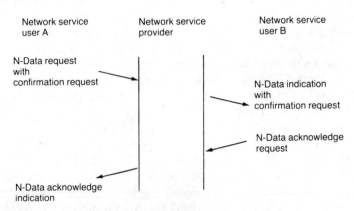

Fig. 4.12 Confirmation of receipt.

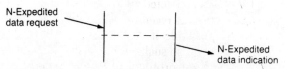

Fig. 4.13 Expedited data transfer.

The sequence of primitives is illustrated in Fig. 4.13. There are no responses or confirmations for Expedited. Protocol mechanisms may, and invariably do, limit the number of Expedited data units outstanding at any given time (via a protocol acknowledgement scheme). The mapping of these on to the layer service interface mechanisms is a local implementation matter.

Although X.25 originally (and up to 1980) had the Expedited feature by means of the Interrupt packet, the amount of data was limited to one octet. Only in the 1984 version of X.25 is 32 octets permitted, and the required negotiation mechanism included.

4.4.1.7 Reset

The reset service is a mandatory service and the possible sequences of primitives is shown in Fig. 4.14(a) to (d).

The Reset request/indication primitives have originator and reason parameters, identical in usage to those described for the Disconnect primitives. The destructive property of Reset will cause the discard within the Network service povider (for both directions of transmission) of any undelivered:

(a) NSDUs;

(b) (N)-expedited data units;

(c) Confirmations of receipt.

The resynchronising (and 'flushing') property is defined such that it is guaranteed that any items received after a reset cannot be items that were submitted at the remote end before the reset occurred. In this context an item may be a NSDU, an Expedited data unit, or a confirmation of receipt. However, this does not necessarily require the (N)-Reset confirm to be delayed until the remote end has completed the reset procedure, as indicated by the tilde (˜) symbols in the figures.

Note: the CCITT network service definition X.213 allows the originator parameter to take a value of undefined. Thus, in practice, implementors are well advised to take account of this.

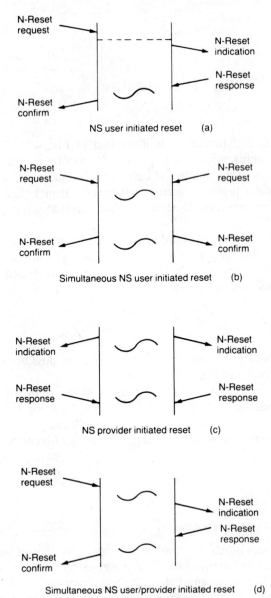

Fig. 4.14 Reset.

4.4.1.8. Semi-formal definition of Network service

In addition to the primitive sequence diagrams the service is defined semi-formally with the aid of a queue model and a state transition diagram. The

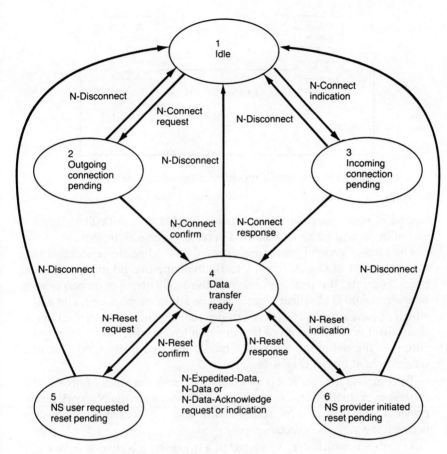

Fig. 4.15 State transition diagram for sequences of primitives at an NC endpoint.

state transition diagram is of traditional style and defines the local events across the conceptual layer service boundary, i.e. the local Network connection (NC) endpoint. This is shown in Fig. 4.15.

The queue model attempts to describe (abstractly) the behaviour and effects that take place within the Network Layer. The presence of intermediate systems with authority and autonomy of their own, and their indirect side-effects are difficult to describe except in very general terms. For example, whether an Expedited data transfer 'overtakes' a normal data transfer depends on whether blockage is occurring, and even then on the amount of blockage and resources available. Furthermore, the extent of overtaking may vary for individual intermediate systems depending on their design. Similar considerations apply to Reset and Disconnect in

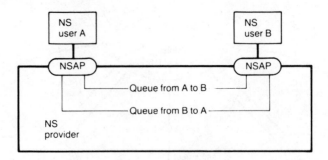

Fig. 4.16 The queue model of a network connection.

respect of their destructiveness. Flow control is also difficult to describe except in general terms as discussed previously in section 3.6.4.

The kind of general properties required, are properties associated with the movement of objects from A to B, involving overtaking, destruction, precedence, etc. It is thus convenient to think of a queue of objects existing between A and B to which objects may be added or extracted, and within which objects may be re-ordered or deleted. Furthermore, certain operations affect both directions of transmission whilst others do not, and so to complete the definition the queue model has two queues, one for each direction as shown in Fig. 4.16.

To complete the model it is necessary to specify the kinds of objects that can be placed in the queue by the Network service user and Network service provider, the flow control relationships, and the ordering relationships between the various objects.

The objects which may be placed in a queue by a Network service user are:

(a) connect objects (representing (N)-Connect primitives and all of their parameters);

(b) octets of normal data;

(c) indications of end-of-NSDU (completion of an (N)-Data primitive);

(d) Expedited NSDUs (representing (N)-Expedited-Data primitives and all their parameters);

(e) data acknowledge objects (representing (N)-Data-Acknowledge primitives);

(f) reset objects (representing (N)-Reset primitives and their parameters);

(g) disconnect objects (representing (N)-Disconnect primitives and all their parameters).

The objects which may be placed in a queue by the Network service provider are:

(h) reset objects (representing (N)-Reset primitives and all their parameters);

(i) synchronisation mark objects;

(j) disconnect objects (representing (N)-Disconnect primitives and all their parameters).

All of the objects above, with three exceptions, are service primitives. Service primitives are normally defined as atomic events as far as the Network service users are concerned. A different level of abstraction is required, however, with respect to normal data transfer and Reset in order to describe the mechanics of the Network service provider.

For normal data transfer, the NSDU size cannot be stated and thus the action of Expedited can only be effective if pieces of data in transit forming part of an arbitrarily long NSDU can be overtaken. It is thus necessary to define octets of normal data, and indication of end-of-NSDU, to explain the behaviour of the Network service provider.

For reset, the 'before' and 'after' effect needs to be described. On initiating a Reset a synchronisation mark is placed in the queue after which all received data will be destroyed until a corresponding Reset object is received. This ensures that once a Reset has begun, no data will be delivered until the provider is sure that the Reset has been completed at the remote end. Therefore, any data received after a Reset could only have been entered after the corresponding remote Reset and not before.

Once in the queue, the Network service provider may manipulate pairs of adjacent objects, resulting in:

(a) change of order; the order of any pair of objects may be reversed, if and only if, the following object is of a type defined to be able to advance ahead of the preceding object. No object is defined to be able to advance ahead of another object of the same type;

(b) deletion; any object may be deleted if, and only if, the following object is defined to be destructive with respect to the preceding object. If necessary, the last object on the queue will be deleted to allow a destructive object to be entered – they may therefore always be added to the queue. Disconnect objects are defined to be destructive with respect to all other objects. Reset objects are defined to be destructive with respect to all other objects except Connect, Disconnect, and other Reset objects.

The entire service and permissible provider behaviour can now be defined in terms of a simple precedence table as shown in Fig. 4.17. For instance, taking the intersection of column 4, row 2 it can be seen that an Expedited NSDU may overtake octets of normal data. Column 8, row 1 indicates that a Disconnect object may destroy a Connect object.

To complete this section on the Network service definition a summary of the primitives and parameters from ISO 8348 is shown as Fig. 4.18.

4.4.2 Provision of the OSI connection-mode Network service (CONS)

This section will deal with the protocols that can be used to provide the CONS over real communication networks/media (subnetworks in OSI terminology).

Although the potential scope is enormous, the scope is limited in practice by the small number of international standards available or under development that can be applied.

Provision of the CONS is simply a matter of taking a particular subnetwork, and comparing its services with those required for OSI. This will result in either the standard protocol of the subnetwork being directly usable, or usable in conjunction with a supplementary 'enhancement' standard. The following types of subnetwork will be considered:

(a) X.25 based packet-switched networks;

(b) local area networks;

(c) X.21-based circuit-switched networks;

(d) PSTN (Public Switched Telephone Network);

(e) ISDN (Integrated Services Digital Network);

(f) general subnetworks providing a connectionless-mode subnetwork service.

Operation over many of these subnetworks may be expected to use standard protocols which do not inherently provide the OSI Network service. In this case, additional standard protocols will be required to increase the level of functionality to correspond to that of the OSI Network service.

In theory, it is the Data Link Layer that provides the service for the basis from which to build the Network Layer. This service will determine what functionality is required within the Network Layer. Thus, it could be argued that we are not yet in a position to describe how to provide the OSI Network service since we have not considered the Data Link service.

In practice, however, it is exceedingly difficult to identify exactly what the Data Link service is in some real communications networks. Generally,

Following object x is defined with respect to preceding object y	Connect	Octets of normal data	End of NSDU	Expedited NSDU	Data acknowledgement	Reset	Synchronisation mark	Disconnect
Connect	—	—	—	—	—	—	—	DES
Octets of normal data	—	—	—	AA	AA	DES	—	DES
End of NSDU	—	—	—	AA	AA	DES	—	DES
Expedited NSDU	—	—	—	—	AA	DES	—	DES
Data acknowledgement	—	—	—	AA	—	DES	—	DES
Reset	—	—	—	—	—	DES	—	DES
Synchronisation mark	—	—	—	—	—	DES	—	DES
Disconnect	—	—	—	—	—	—	—	—

Key: 'AA' indicates that object x is defined to be able to advance ahead of the preceding object y.
  'DES' indicates that object x is defined to be destructive with respect to the preceding object y.
  '—' indicates that object x is neither destructive with respect to object y nor able to advance ahead of object y.

Fig. 4.17 Ordering relationships between queue model objects.

Phase	Service	Primitive	Parameters
NC establish	NC establish	(N)-Connect request	(Called address, calling address, receipt confirmation selection, Expedited data selection, QOS parameter set, NS user-data)
		N-Connect indication	(Called address, calling address, receipt confirmation selection, Expedited data selection, QOS parameter set, NS user-data)
		(N)-Connect response	(Responding address, receipt confirmation selection, Expedited data selection, QOS parameter set, NS user-data)
		(N)-Connect confirm	(Responding address, receipt confirmation selection, Expedited data selection, QOS parameter set, NS user-data)
Data transfer	Normal data transfer	(N)-Data request	(NS user-data, confirmation request)
		(N)-Data indication	(NS User-data, confirmation request)
	Receipt confirmation (see note)	(N)-Data-acknowledge request	—
		(N)-Data-acknowledge indication	—
	Expedited data transfer (see note)	(N)-Expedited-Data request	(NS user-data)
		(N)-Expedited-Data indication	(NS user-data)
	Reset	(N)-Reset request	(Originator, reason)
		(N)-Reset indication	(Originator, reason)
		(N)-Reset response	—
		(N)-Reset confirm	—
NC Release	NC Release	(N)-Disconnect request	(Originator, reason, NS user-data), responding address
		(N)-Disconnect indication	(Originator, reason, NS user-data, responding address)

Note: NS provider option: may not be provided in every Network service.

Fig. 4.18 Summary of network service primitives and parameters.

there is either some set of protocols or perhaps one monolithic protocol that inherently make up the subnetwork interface. These may not provide a 'clean' separation between the Data Link and Network Layers. Even where some kind of separate Data Link protocol is apparent, there are large variations from one subnetwork to another. For example, the use of an X.21-based circuit switched interface is very different from the use of an X.25-based packet-switched interface in respect of Data Link operations, and different again from local area network operation.

The work of defining the Data Link service itself therefore has been complicated by the above considerations, with philosophies ranging from arranging a multiplicity of Data Link services to reflect real subnetworks, to the definition of a single service for future subnetworks (and in consequence different from existing ones). Even now, the Data Link service has not been agreed.

In practice, this does not matter but it means that one has to examine the total functionality of subnetworks individually and base provision of the OSI Network service on the usable subset of their functionality. There is no necessity to break down this functionality to determine a discrete Data Link service. It is to be noted that similar considerations apply to the Physical Layer service, where again subnetworks offer their own variety of interface and associated protocols (e.g. modem interfaces, LAN interfaces, ISDN interfaces, etc.).

In summary, then, one has to examine particular subnetworks as a whole in order to assess how they may be employed to provide the Network service. However, work is proceeding on the definition of the Data Link service and this will be discussed in section 4.8. As far as practical implementations are concerned it will make little difference, since one merely relabels the subnetwork protocol elements in accordance with the ultimate Data Link service definition.

4.4.2.1. Provision of OSI Network service using the X.25 protocol

CCITT first published the Recommendation X.25 for packet-switched networks in 1976. Since that date two major revisions have taken place, one in 1980 and the other in 1984. Most networks in the world today provide at least the 1980 flavour, and are committed to implementing the 1984 version. However, the 1980 version does not support all the functionality required for the OSI Network service, and so ISO has published the standard ISO 8878 containing an enhancement protocol (an SNDCP). ISO 8878 also (and primarily) contains a specification of how to use the 1984 version of X.25 to provide the OSI Network service. The latter is necessary because X.25 contains a number of facilities and features upon which usage guidelines are required. It must be realised that X.25 may be used in other ways when not being employed specifically to provide the Network service. A generic usage document for X.25 is published by ISO at IS 8208, and is an extremely useful companion document to the CCITT Recommendation X.25 from a terminal implementor's perspective.

The CCITT Recommendations are always defined in terms of the procedures across the interconnection interface as perceived by the subnetwork provider (called the data circuit-terminating equipment, (DCE) in CCITT terminology). The ISO standards on the other hand, are defined in terms of the procedures across the same interface but from the subnetwork-user perspective (termed the data terminal equipment, (DTE) by CCITT).

This section will assume that readers are either familiar with X.25 or will refer to the appropriate references. X.25 will not be described *per se*, but only in relation to the OSI Network service.

There are two circumstances in which use of X.25 may arise:

Case (a) communication across a subnetwork to which the access protocol is X.25 (a real X.25-based packet-switched network, e.g. UK PSS, USA – Telenet, Canada – Datapac, etc.);

Case (b) communication across non-X.25 subnetworks over which X.25 is being used as an enhancement protocol (for use over LANs, or even over circuit switched networks).

Case (a) is relatively straightforward and will be described first, since the Packet Level protocol (PLP) of X.25 is also required for case (b). For case (a) the whole of X.25, i.e. both the Data Link Level and the Packet Level are as given. A detailed analysis of the split between the Data Link Level and the Packet Level is not required. It is sufficient to know that the combined functionality of the two levels meets (or in one case very nearly) the functionality requirements of the OSI Network service. For case (a) the Data Link Level provides the necessary sequencing/resequencing mechanisms to provide to the Packet Level a guaranteed in-sequence 'data pipe' even in the face of transmission errors. The potential division between the Packet Level and the Data Link Level will require further analysis for case (b), where choices exist for the Data Link Level protocols. ISO has produced a standard ISO 7776 for the X.25 LAPB compatible Data Link procedures, which will be discussed later.

4.4.2.2 Use of an inherent X.25 subnetwork 1984 version

A subnetwork which operates in accordance with the 1984 version of X.25 requires a minimum of extra protocol in order to provide the CONS. In fact, no real pure new protocol elements are required. It is only necessary to specify usage of, and parameters for, the inherent protocol elements of X.25. ISO has produced standard ISO 8878 specifying how X.25 Packet Level Protocol (PLP) is to be used to provide the CONS, based on reference to the totality of more general procedures already specified in ISO 8208 (which corresponds to the CCITT Recommendation X.25 itself).

X.25 facilities fall into two basic categories: those that are invoked by request (on demand), and those that specifically have to be subscribed to (for a period). The latter determine the receiver's capabilities in relation to the initiator's requests. Thus ISO 8878 specifies both subscription and usage (request) requirements. Some facilities are not of a subscription nature even though they may be optionally requested.

CCITT, itself, also defines its facilities into 'E' or 'A' classifications. Facilities with an 'E' (Essential) classification must be made available by all subnetwork providers. Facilities in the 'A' (Additional) category may only be available in certain subnetworks. 'E' or 'A' facilities may themselves be further qualified by the attributes mentioned in the previous paragraph.

In addition to the facilities defined by CCITT applicable to the X.25 subnetwork itself, X.25 makes provision for user-defined facilities. ISO has collaborated on (a) the provision of this capability; and (b) encoding of these fields to prevent encoding clashes and to assist possible migration by CCITT for subnetwork usage of such elements at a future date. For these reasons these facilities are termed 'CCITT-specified DTE facilities', and are conveyed transparently across the X.25 subnetwork.

The provision of the ISO CONS requires the following X.25 optional-user facilities and CCITT-specified DTE facilities to be invoked and or subscribed to:

(a) Optional-user facilities
   Fast select (facility to be invoked)
   Fast select acceptance (facility to be subscribed)
   Throughput class negotiation (both to be invoked and subscribed to)
   Transit delay selection and indication (facility to be invoked);

(b) CCITT-specified DTE facilities
   called address extension
   calling address extension
   End-to-end transit delay negotiation
   Expedited data negotiation
   Minimum throughput class negotiation.

A summary of mappings for the Network connection establishment phase is given in ISO 8878. Further parameters of X.25 may be used in conjunction with those required to provide the OSI CONS, e.g. reverse charging, flow control parameters, etc.

The receipt confirmation negotiation is achieved not through facility requests, but through the setting of bit 7 of the General Format Identifier (GFI) in octet 1 of the X.25 packets.

The selection of Expedited data follows the same rules but mechanised via the CCITT-specified DTE facility.

4.4.2.2.1 THROUGHPUT QOS PARAMETERS

For (N)-Connect request and indication primitives the Network service 'target' throughput parameter is mapped to the X.25 throughput class negotiation optional user facility, and the 'lowest acceptable' is mapped to the minimum throughput class negotiation CCITT-specified DTE facility of X.25 call request/incoming call packets. As the Network Connection establishment progresses, Network Layer entities may reduce the target value according to its available resources. If a Network Layer entity receives an incoming call packet with the target value less than the lowest acceptable the X.25 call must be cleared with an appropriate X.25-defined diagnostic which will map to the (N)-Disconnect reason-parameter. Diagnostic codes 229 and 230 have been assigned to 'QOS-not-available transient condition' and 'QOS-not-available permanent condition' respectively. The selected value parameters of (N)-Connect response/confirm is mapped to the throughput class negotiation facility of the call accepted/connected packets.

4.4.2.2.2 TRANSIT DELAY QOS PARAMETERS

The handling of transit delay is more complicated than for throughput, since it involves more related parameters and some processing.

X.25 itself defines a Transit Delay Selection and Indication facility (optional user facility) and additionally the CCITT-specified DTE facility defines three other parameters for: (a) cumulative transit delay, (b) target transit delay; and (c) maximum end-to-end transit delay. The latter two parameters correspond to the Network service parameter of the (N)-Connect request and indication primitives. The other two parameters are required to take account of actual delays incurred by the X.25 subnetwork and any other delay components attributable to other subnetworks in the chain or within the end-systems.

The X.25 parameter transit delay selections and indication is itself a little complicated. The idea is that the Network Layer entity receiving the incoming call packet should at least receive an indication of the transit delay component attributable to the X.25 subnetwork. Additionally, some networks may offer choices of transit delay and thus in this case, the required transit delay could be requested in the call request packet. The X.25 subnetwork would then invoke whatever measures are necessary to meet this request. In the event that the subnetwork does not offer any choices the subnetwork itself merely inserts its own estimate of the transit delay attributable to itself.

Network Layer entities which receive an incoming call packet at intermediate or end-system have to perform certain computations.

When dealing with an (N)-Connect request the Network service provider local entity maps the target and maximum acceptable values to the CCITT-specified DTE facilities, target and maximum acceptable. In addition, the Network Layer entity inserts in the cumulative transit delay parameter the amount of transit delay attributable to the calling end-system, and selects a value for the X.25 subnetwork transit delay (if a choice is available) by using the transit delay selection and indication facility.

Any Network Layer entity receiving an incoming call packet has to compute the total cumulative transit delay up to and including itself. It does this by replacing the current received cumulative transit delay with the summation of the values of the transit delay selection and indication facility, the current received cumulative transit delay, and its own transit delay. If at any location (intermediate or end-system) this value exceeds the maximum acceptable the call is cleared. Again, diagnostics 229 and 230 are used as appropriate.

4.4.2.2.3 DATA TRANSFER SERVICE

The OSI Network service data units (NSDUs) are provided by using the X.25 packet-sequencing mechanism. Thus the N-Data request/indication corresponds to a sequence of one or more X.25 data packets (depending on the size of the NSDU) where all data packets except the last have the X.25 (M)-bit set to 1, and the (D)-bit set to 0. The last packet will have the (M)-bit set to 0, and the setting of the (D)-bit will depend on whether the (N)-Data request primitive has the confirmation request parameter set.

If the (N)-Data request has the confirmation request parameter set, then the (D)-bit of the last packet of the (M)-bit sequence of packets must also be set to 1. The 'sink' Network Layer entity receiving an X.25 data packet with the (D)-bit set, must not transit the P(R) acknowledging receipt of that packet until it has received the corresponding acknowledgement from the local Transport entity. This is achieved via the N-Data acknowledge request primitive of the OSI Network service. The precise mechanisation of this is a local implementation matter.

When the sink Network Layer entity receives the N-Data acknowledge request, the appropriate P(R) is generated and transmitted to the source Network Layer entity using the appropriate X.25 packets, e.g. RR, or an available data packet. The source will interpret the received P(R) as the N-Data acknowledge indication, corresponding to the data packet it had sent with the (D)-bit set to 1, and signal this primitive to the Network service user (Transport entity).

4.4.2.2.4 EXPEDITED DATA TRANSFER SERVICE

The Network service requirements for Expedited data transfer are easily accommodated by X.25.

The service itself is only available during the data transfer phase if its use has been agreed during connection establishment via the option negotiation mechanism.

The N-Expedited data request primitive maps directly to the X.25 interrupt packet which has a user data field capable of accommodating up to 32 octets of user data. At the receiving end the X.25 packet is mapped to an N-Expedited data indication. X.25 procedures restrict the number of outstanding (unconfirmed) Interrupt packets to one (i.e. a second packet cannot be sent until the previous one has been acknowledged). How the restriction affects the local service boundary in terms of interface flow control, and maintenance of the relationship required between Normal and Expedited data is a weak implementation matter.

4.4.2.2.5 RESET SERVICE

The mapping of N-Reset Request to the X.25 Reset Request and N-Reset Indication to X.25 Reset indication packets is straightforward.

The potential relative differences of timings between the X.25 Reset confirmation packet and the Network service user-generated N-Reset Response, and the N-Reset Confirm primitives gives rise to several different situations and possible design choices. The coupling between the service primitives and the X.25 Reset confirmation packet may be very direct or indirect as a result. Furthermore, the design may be different at each end in this respect. The overriding criterion is that the 'before' and 'after' effects of reset must be preserved. This was discussed in section 4.4.1.

In order to preserve the 'before' and 'after' effects the following criteria are applied:

(a) when an N-Reset response primitive is received the Network Layer entity must accept N-Data and N-Expedited data request, primitives from the Network service users for subsequent transmission after completion of the X.25 PLP Reset procedure (if not already completed);

(b) when the Network Layer entity signals the N-Reset confirm primitive to the Network service user, it must accept N-Data and N-Expedited data primitives from the Network service user, for subsequent transmission after completion of the X.25 PLP Reset procedure (if not already completed);

(c) when an N-Reset Indicator has been signalled, the Network Layer entity will discard all primitives until an N-Reset response is received;

(d) when an X.25 Reset request has been transmitted, the Network Layer entity will discard all X.25 packets until the X.25 reset has been completed.

The combination of X.25 cause and diagnostic code fields are mapped to the combination of the N-Reset primitives parameters, originator and reason. ISO 8208 defines a large range of diagnostics, over and above those defined in the Network service definitions. This results in the many X.25 diagnostics being mapped, but only two reason parameters; congestion, and 'reason unspecified'. This loss of information at the service boundary could be resolved by logging the actual X.25 diagnostic codes for use by the layer management entity.

4.4.2.2.6 CONNECTION RELEASE

An N-Disconnect request received from a Network service user is mapped to an X.25 clear request packet, by the Network Layer entity.

When an X.25 clear indication packet is received, the Network Layer entity transmits an N-Disconnect indication primitive to the Network service user, and transmits an X.25 clear confirmation packet.

A Network Layer entity itself may wish to release the Network connection and may do so by merely transmitting an X.25 clear request, and an N-Disconnect indication primitive to the Network service user.

The above events are summarised in Fig. 4.19. It can be seen that X.25 clear confirmation packets are not mappable to any Network service primitive. As far as the Network service user is concerned an N-Disconnect request is assumed to have immediate effect. Implementations may use local conventions, if they are to include control of establishment of new Network connections in relation to the confirmed release of the X.25 connection. It should also be noted that the significance of an X.25 clear confirmation packet may only be local anyway, and thus not indicate the state of the remote Network service provider/user. Furthermore, a collision of Disconnect primitives across the Network service boundary is not possible by definition, unlike those across the X.25 DTE/DCE interface which are handled in accordance with defined procedures (ISO 8208/X.25).

4.4.2.2.7 X.25 RESTART

Restart is not directly equivalent to any Network service primitive. The OSI Network service is specified for an instance of communication (i.e. a single connection between a pair of NSAPs). Thus, a given logical channel represents an instance of the Network service. The multiple simultaneous

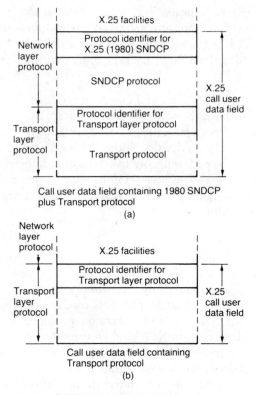

Fig. 4.19 Protocol identification.

calls capability of X.25 by means of different logical channels provides multiple instances of the Network service and a logical channel number is synonymous with the OSI connection-endpoint-identifier.

Since the restart procedure affects all instances of communication, the only possible mapping for a received restart indication packet is to generate a number of N-Disconnect indications, one for each affected logical channel. The cause code and diagnostic code fields are treated as for clear indication packets.

There is no equivalent to the restart request. The generation of an X.25 restart request can only be regarded as a local management function. There is nothing to prevent local implementations making local arrangements to handle the restart procedures more directly.

4.4.2.2.8 OTHER X.25 FEATURES

There are many other X.25 features which may be employed, quite freely, since they do not impact the OSI Network service, e.g. one-way logical channels, closed user groups, reverse charging, etc. The OSI Network service will permit use of any of the remaining X.25 facilities not previously discussed with the exception of the (D)-bit modification facility. This must never be employed since it is contrary to the principle of negotiating the use of receipt confirmation.

4.4.2.3 Use of an inherent X.25 subnetwork 1980 version

Many changes were made to X.25 during the 1980–84 CCITT Study Period. Recognising that it would take several years for real X.25 networks (subnetworks in OSI terminology) to be upgraded, ISO has produced a 'thin' subnetwork dependent convergence protocol (SNDCP) for use with this pre-OSI version of X.25.

With the exception of the expedited data transfer all elements of the OSI Network service are accommodated.

The SNDCP specified in ISO 8878 uses X.25 fast select if available, although it is not essential. The basic concept behind the SNDCP is to use more than one packet where necessary and to thus regard completion of a Network service primitive as a possible sequence of packets rather than the one-to-one correspondence available from X.25 (1984).

The extra parameters required during call set-up are placed in X.25 data fields since there is no provision for the 1984 CCITT-specified DTE facilities. The codings chosen, however, are identical to those specified in X.25 (1984), thus aiding migration.

The basis of the X.25 (1980) SNDCP is to use the transparent data fields of X.25 to carry the extra elements of Network Layer protocol necessary to fulfil the OSI Network service. This creates a need to distinguish and delineate between Network Layer protocol and Transport Layer protocol, both of which share the X.25 data fields. This demarcation is achieved in two ways. For call establishment packets the first octet of the X.25 call user data field is used to indicate whether the second octet contains Network Layer protocol or Transport Layer protocol, and is termed a 'protocol identifier'.

At this point, it is necessary to explain this protocol identification scheme in more detail since it has important consequences and ramifications both for OSI and non-OSI usage and combinations of OSI/non-OSI usage.

The protocol identifier acts as a multi-purpose identifier, *essentially providing information* on how X-25 is being used. This is especially useful since each individual X.25 call within a set of simultaneous calls (to a given address) may be carrying different combinations of higher level protocols. This potential capability was first introduced into CCITT Recommendation X.29 which uses X.25 in a special way for passing control information from a packet-mode DTE to a Packet Assembly/Disassembly Facility

(PAD). A special Recommendation X.244 has recently been introduced to emphasise the multi-purpose nature of this feature and to co-ordinate allocation of encodings. It is essential to ensure that the encoding of this field is unique across all the layer, since one of its uses is to detect to which layer a protocol belongs.

For example, take the case where the OSI Transport protocol is being used: (a) with 1980 SNDCP; and (b) with 1984 X.25. Case (a) is illustrated in Fig. 4.19 (a). For case (b), Fig. 4.19 (b) would apply. In order to distinguish case (a) from case (b) it is clear that the encoding for the Network Layer protocol identifier must be different from that of the Transport Layer protocol identifier even though they are by definition operating in different layers. X.244 sets out the principles for use of the protocol identifier, and the current coding allocations are shown in Fig. 4.20. Bits 7 and 8 are effectively used to distinguish between CCITT, ISO, national, and private protocols. It should be noted that what follows the protocol identifier is dependent upon the value of the protocol identifier, e.g. nothing for Teletex, three octets for X.29, etc., in accordance with the appropriate standard. The protocol identifier can be regarded as a recommendation/standard identifier.

For data packets the Q-bit is used to distinguish Network Layer protocol from Transport Layer protocol. In the particular case where the combination of extra parameters and the Network service user data will not fit into a fast select packet or when fast select is not available the OSI Network connection establishment requires the combination of X.25 call set-up and data packets. This is achieved by placing a continuation parameter in the Call User Data field, and then conveying the extra information in subsequent data packets.

8	7	6	5	4	3	2	1	protocol identifier
0	0	← CCITT-defined ——————→						
0	1	← nationally defined ————→						
1	0	← ISO-defined ——————→						
1	1	← privately defined ———→						

CCITT-defined

8	7	6	5	4	3	2	1	
0	0	0	0	0	0	0	1	X.29
0	0	0	0	0	0	1	0	Teletex
		see section 5.2						transport protocol

ISO-defined

8	7	6	5	4	3	2	1	
1	0	0	0	0	1	0	0	1980 X.25 SNCDP

Fig. 4.20 Protocol identifier encodings.

In practice, when fast select is available this 'overflow' in subsequent data packets will not be necessary since there is no OSI (or even non-OSI) standard which requires a Network service user data field of more than a few octets.

Provided no Network service-user data is required during the connection release, the standard 1980 X.25 clear packets may be used. In practice, this is always the case since there is no current requirement specified in OSI for use of the Network service-user data parameter. The parameter itself, however, is defined as being available in the Network Service and thus a procedure using X.25 data packets is defined in 1980 X.25 SNDCP. In this procedure the Network connection would be released by use of X.25 data packets (with Q bit set, etc.) prior to the X.25 proper release phase.

One other minor difference is the use of the diagnostic field of X.25 clear and reset packets. X.25 (1980) does not permit use of the cause field by the DTE. Thus, a DTE acting as a Network Layer relay, must use the diagnostic field to indicate a provider initiated event.

As far as the SNDCP protocol encoding is concerned the parameters have the same encoding as for X.25 (1984). It is thus only necessary to have additional non-X.25 encodings for message types (e.g. connect request, disconnect request, etc).

ISO 8878 defines the generic encoding for X.25 data packets for carrying the SNDCP. Message codes and parameter types are encoded in the style of X.25, for 1-octet, 2-octet, 3-octet, and variable-length parameter value fields (i.e. as for X.25 class A, B, C, D). Where there is an equivalent X.25 (1984) facility the same coding is used. When there is no direct X.25 equivalent, bit 6 of field is set to 1 to ensure no conflict with any other CCITT-defined facility.

Finally, it should be noted that use of X.25 (1980) SNDCP does not provide the OSI Expedited data transfer. This is because to be effective it must be available from the subnetwork itself, it cannot simply be 'grafted' on. For example, the transmission of 32 separate X.25 Interrupt packets would not suffice nor would simple transmission of data packets with the Q-bit set since the by-passing of flow-controlled data could not be effected.

4.4.2.4 Use of X.25 over leased lines and circuit switched subnetworks

These are situations in which it is required to realise the OSI CONS over either a dedicated point-to-point leased line or where a circuit switched network is used to effect a point-to-point connection for the Physical Layer (to access a packet switched network, for example), (see Fig. 4.21).

These situations may arise between OSI end-systems, between OSI end-systems and intermediate systems, or between intermediate systems. In

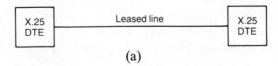

(a)

Fig. 4.21(a) Use of X.25 over leased line.

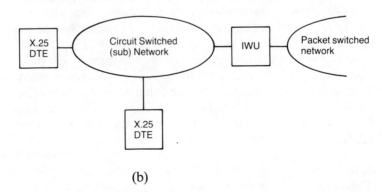

(b)

Fig. 4.21(b) Use of X.25 over circuit switched subnetwork.

these cases, X.25 is invoked subsequent to the establishment of the point-to-point connection. A circuit-switched connection may be of any type, e.g. PSTN, or X.25 can be employed to advantage in these situations with the addition of a small number of specific implementation requirements, over and above those previously specified.

Primarily, the problem that arises out of the use of X.25 is how to deal with the lack of symmetry and *a priori* parameterisation normally associated with X.25. For example, many of the X.25 procedures are based on an agreement between the two communicating entities as to who is to act as a DCE and who is the DTE. For example, this is important for the Packet Level in order to maintain the logical channel assignment rules (a DTE selects logical channels starting at the high order, a DCE selects channels from the low order end of the range of available channel numbers).

ISO 8208 section 4.5 describes how the restart procedure is used to determine this. Basically the procedure permits either Network Layer entity to initiate the restart procedure dictates the outcome. The possible simultaneous collision possibility is resolved by detection of collision and subsequent re-attempt after a random time delay.

Some additional conventions are also required for the Link Level of X.25, to resolve Link Level address assignment.

4.4.2.5 Provision of CONS over Local Area Networks

Three methods have been standardised in ISO 8881 for the provision of the CONS over local area networks (LANs). Both methods involve use of X.25. Choice of method depends on the exact LAN configuration under consideration, complicated by the availability of various flavours of Data Link Layer standards for LAN operation. Basically there are two flavours of Data Link Layer protocols, one providing the connectionless service, the other connection-orientated. However, the connectionless flavour is not pure connectionless, since it preserves sequence such that mis-ordering is not possible, across a single LAN, but losses can occur due to transmission error.

For the latter case, since a resequencing protocol is unnecessary it would be possible to mount the Packet Level protocol (PLP) of X.25 directly on top of this Data Link Layer protocol, provided that losses can be detected and recovered. However, X.25 (PLP) was not initially designed for this purpose and requires the addition of timers and retransmission procedures to be effective.

However, for the single LAN case, if the loss rate is high, a further alternative is the use of connection-orientated Data Link Layer protocol instead (equivalent to X.25 LAPB) over which X.25 PLP can then be operated efficiently.

The ISO LAN Data Link Layer protocols providing the connectionless and connection-orientated services are defined in ISO 8802: Part 2 by logical link control procedures, type 1 (LLC1) and Type 2 (LLC2) respectively. The various protocol combinations are shown diagrammatically in Fig. 4.22.

4.4.2.5.1 USE OF X.25 OVER LLC2 (ISO 8881)

In addition to the Network service requirements specified in ISO 8878, the X.25 Packet Level protocol (PLP) has to be operated in the DTE-to-DTE mode specified in ISO 8208 (X.25 Packet Level protocol for data terminal

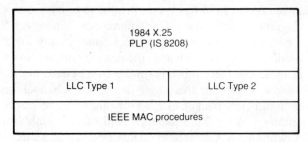

Fig. 4.22 Protocols to provide CONs over LANs.

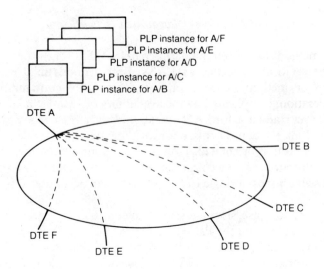

Fig. 4.23 Use of Multiple links for X.25 LAN operation.

equipment). For LAN operation, a Data Link Service has to be established between each pair of stations on the LAN that wish to intercommunicate. Thus, if a given LAN station wishes to communicate with, say, five other stations, five separate Data Links will be in operation, and consequently these will be five separate instances of the X.25 Packet Level protocol at that given station (see Fig. 4.23).

The management of logical channels becomes more complex in the case of LAN operation due to the existence of multiple instances of the X.25 Packet Level Protocol per station. A simple solution would be to standardise the number and range of logical channels per X.25 instance per station. This, however, does not readily take account of individual station needs, and could be onerous to some stations and too limiting for others. A more complicated alternative is to have pair-wise bilateral agreements between stations, which would require significant amounts of recorded information and the update/management problem.

An enhancement to X.25 has been introduced to overcome the logical channel assignment difficulties. This enhancement involves the use of source and destination references to be individually assigned by both the calling and called DTE. This obviates the need for any *a priori* knowledge between stations. The mechanism is similar to that used in the Transport Layer protocols, except in this case, the logical channel field of call establishment and clear packets is used for one of the references, and the other is contained in the X.25 facilities field of call establishment/clear packets, by defining an X.25 type-B facility designated 'source reference'.

These are two variants of the enhancement, depending on whether the

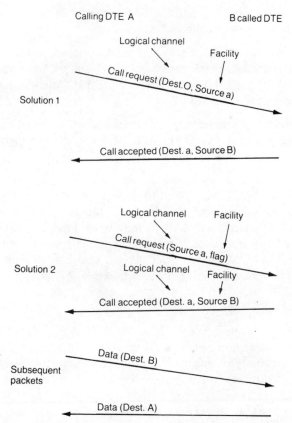

Fig. 4.24 X.25 source/destination reference enhancement.

calling or called DTE uses the logical channel field, or the source reference facility field for its source reference. The two schemes are shown in Fig. 4.24. Solution 2 requires a type-A facility in the call Request to indicate use of the enhancement procedure, since it cannot be otherwise deduced at the called DTE. In solution 1 the use of the enhancement is obvious, from use of the facility field.

For subsequent packets transmitted by a DTE the logical channel number is set to the value of the destination reference, i.e. the source reference of the remote DTE. It can be seen that this results in use of two logical channel numbers after a call has been established, one for each direction of transmission. The clearing procedures also include the reference mechanism.

As far as addressing is concerned the media access control (MAC) and logical link control (LLC) addresses defined in ISO 8802/2 to 6 are utilised as follows:

(a) both the MAC source and destination address fields correspond to subnetwork points of attachment (SNPA);

(b) the combination of MAC source and destination addresses uniquely identifies the Packet Level entity pertaining to the corresponding instance of the DTE/DTE interface (recalling that these will in general be multiple instances per DTE);

(c) the LLC source and destination addresses identify use of the X.25 Packet Level protocol. This can enable distinction to be made between usage of LLC links between the same pair of DTEs.

The question of link management is important for LLC 2 operation. Clearly, we cannot avoid having point-to-multipoint operation. However, since link set-up collisions can occur between stations it is important to ensure that only the necessary required number of links are established between a given pair of stations. Such collisions can be avoided by only allocating one LLC address for X.25 usage. Collision of link set-up commands can then be resolved in the usual way.

4.4.2.5.2 USE OF X.25 OVER LLC1 (ISO 8878)

LLC1 provides a connectionless Data Link service. Unlike a pure connectionless service, however, operation over a single LAN does not introduce the possibility of misordering, nor can duplication occur. The MAC layer detects frame errors (transmission errors) and discards frames found to contain errors. Thus occasionally an X.25 packet may be lost.

If the X.25 PLP is to be operated over LLC1, account needs to be taken of possible packet loss in the X.25 PLP procedure itself. Primarily this requires specific use of the retransmission procedures and timers specified in ISO 8208. In addition, the relative magnitudes of the related timers becomes extremely important if deadlocks are to be prevented.

The main problem concerns loss of a data packet or loss of a packet containing an acknowledge (a new P(R)). It should be noted that the loss of a data packet can result in the next data packet being considered as an out-of-sequence data packet since the P(S) will not be the next expected. In order to devise a simple and adequate strategy, it is necessary to analyse several cases, as shown in Fig. 4.25. The strategy must be capable of working with the various combinations of events that could occur. Figures 4.25 (a) and (b) shows possible transmitter options of sending a reset or retransmitting the data packet (using timer T25 as defined in ISO 8208). In Fig. 4.25 (a) a problem will arise if the reset is lost or the reset confirmation is lost. Figure 4.25 (b) only really works if the packet that was lost was the last within the window (otherwise this results in out-of-sequence packets). Figure 4.25 (c) shows how mis-sequencing can occur, and this becomes

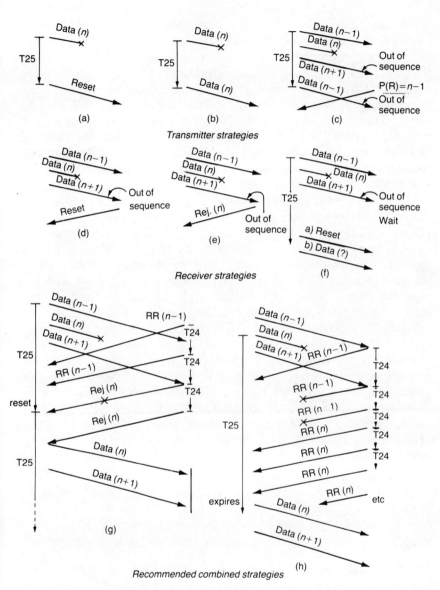

Fig. 4.25 Handling packet loss over LLC1.

complex according to the rate at which P(R)s are returned and whether they in turn themselves are lost. Figure 4.25 (d) and (e) show two possible receiver actions for out-of-sequence packets, i.e. receiving $(n + 1)$ after $(n - 1)$ because n has been lost. If the receiver knows that the transmitter is employing 4.25 (a) or (b) it could just wait.

It is clear from Fig. 4.25 (c) and (f) that the transmitter strategy alone is insufficient since in the event of lost returning P(R)s, be they in RR, RNR, or REJ, the transmitter cannot know precisely which data packet requires retransmission. It therefore could be advantageous if the receiver regularly returned P(R)s also under time-out. It is also desirable to avoid unnecessary resets since they are destructive and lead to further loss of data which would require Transport Layer recovery. ISO 8208 defines a timer T24 for the transmission of P(R)s. Thus, a suitable combination of transmitter's use of T25 and receiver's use of T24 could be used to advantage. Furthermore, this strategy would operate irrespective of whether REJ were being used, if all receivers simply ignored out-of-sequence packets.

The recommended strategy, therefore, is for transmitters to retransmit using T25, and for receivers to ignore out-of-sequence packets and transmit P(R)s using T24. The relative magnitudes of T25 and T24 are important. In this scheme, T25 is required to be much greater than T24. Additionally, receivers could optionally transmit REJ on detection of the out-of-sequence packet. REJ will effect a faster recovery. If REJ is not used, the procedure ensures (assuming that at least one P(R) is received with T25, the transmitter begins retransmission of the next packet expected by the receiver. The procedures for use of T24 and T25 are specified in ISO 8208. Figure 4.25 (g) and (h) show some example sequences using these strategies.

When initialising an X.25 interface it is usual to invoke the restart procedure subsequent to the establishment of the Link Level. Since, in this particular LAN environment, the Data Link service is connectionless, some other criteria must be used to initialise the Packet Level. ISO 8878 recommends that a count of establishing actual calls be maintained. Prior to establishing a new virtual call, the count is examined. If the count is zero then the restart procedure is initiated.

4.4.2.6 Provision of CONS using X.21

CCITT Recommendation X.21 (1984) does not contain all the features required to support the OSI Network service.

The current working document within ISO (SC6 N3150) contains two different approaches to the provision of CONS over an X.21 subnetwork:

(a) use of X.25 after establishing an X.21 connection;

(b) use of an enhancement protocol based on and an extension to CCITT Recommendation T.70.

The first approach needs no further explanation since the previous considerations to using X.25 apply, i.e. use of ISO 8208 and ISO 8878 (how to use ISO 8208 to provide the CONS).

The second approach takes its basis from the procedures developed for Teletex which includes an enhancement protocol to ensure interworking between X.21-based Teletex service and an X.25-based Teletex service as specified in CCITT Recommendation T.70.

As far as X.21 itself is concerned, it is deficient in the following areas in its capability of supporting the OSI Network service:

(a) no capability to carry OSI Network service (NS) Quality of Service (QOS), NSAP address, or Network service user data parameters for Network connection establishment;

(b) no capability for basic features of data transfer phase of the OSI Network service, i.e. no reset, Expedited, or flow control

(c) no capability to carry originator/reason, Network service user data parameters during Network connection release.

During the 1980–84 CCITT study period, provision was made for what is termed subaddressing in X.21. This enables extra information to be transmitted to and from a called DTE during call establishment. This principle, if available at both the calling and called DTEs together with the necessary encoding scheme, could be used as the basis for an extension to X.21 to make it capable of supporting the OSI Network service, for connection establishment.

CCITT Recommendation T.70 specifies the Data Link Layer protocols of X.25 LAPB for use during the data transfer phase of a circuit-switched connection. Additionally, T.70 specifies a minimal network layer (small 'n') protocol during the data transfer phase. This minimal network layer protocol comprises a two-octet network block header, as shown in Fig. 4.26. This header comprises a one-octet length indicator, followed by a network block type code. Only the one block type is currently defined and this contains an (M)-bit and a (Q)-bit (similar to those defined for X.25).

The (M)-bit is used to preserve the integrity of TPDUs and the (Q)-bit is used to provide for functional mapping with the X.25 (Q)-bit for CSPDN/PSPDN interworking. However, it is not used within Teletex itself.

In the same way as the X.25 (1980) SNDCP utilised the (Q)-bit of X.25 data packets, it would be possible to use this procedure within the X.21 data transfer phase. Thus, the current proposal within ISO is to use this technique (i.e. the two-octet network block header) in conjunction with the X.25 (1980) SNDCP. This means that all the formats already defined for X.25 SNDCP are applicable simply by preceding them by the two-octet network block header (see Fig. 4.27). Since the X.21 procedures do not provide the OSI Network service the X.21 procedures are only used to establish and maintain a transparent 'bit pipe' over which the Data Link and Network Layer protocols can be subsequently operated. The qualifier bit (Q) is set to 1 when Network Layer control information is transferred in the Network service user data field. In all other cases (i.e. when transferring Network service user data), it is set to 0.

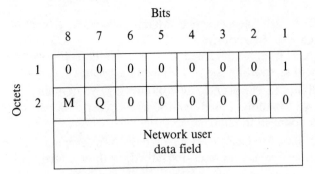

Fig. 4.26 Teletex network block header.

Network Layer Control Information

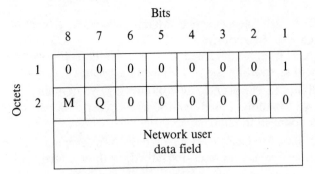

Fig. 4.27 X.21 SNDCP formats.

The first octet in the Network service user data field, when Q = 1 is the message code (MC). The MC identifies the primitive e.g. Connect request, etc. in accordance with the encodings for the X.25 (1980) SNDCP.

Thus, the X.21 SNDCP principles and formats are identical to the X.25 (1980) SNDCP except that receipt confirmation is not provided.

For the Data Link Layer protocol X.25 LAPB is used, and the terminal initiating the X.21 call takes link address A and the terminal receiving the X.21 call takes link address B.

4.5 THE OSI CONNECTIONLESS-MODE NETWORK SERVICE (CLNS) AND ITS PROVISION

As far as the Network Layer is concerned, the connectionless-mode network service has its origins in datagram-style networks, of which the Advanced Research and Projects Agency (ARPA) network in the USA is a prime example. Such networks rely on the user to maintain integrity and sequence, etc. of data, based on an assumption that it is only the job of the network to ensure that small individual pieces of data find their way from A to B, according to instantaneous dynamic routing choices. Such networks by their very nature place extra burdens on the user, i.e. what is optimum for the network may not be optimum for the user.

Typically, an ARPA-like network can result in data being discarded, misordered, or duplicated. Each unit of data, typically, can take one of a variety of routes through the network, due to the network's dynamic routing capabilities. The fact that each unit of data is self-contained (tagged with all necessary addresses and control information) with no associated feedback information, and that under congestion the only recourse is to discard the units (no flow control) leads to the possibilities of loss, misordering and duplication. ISO has produced ISO 8348 AD1 specifying a connectionless-mode network service to standard such a style of operation.

4.5.1 The OSI connectionless-mode network service (CLNS)

In contrast to a connection an instance of the use of the connectionless-mode service does not have a clearly distinguishable lifetime. In addition, the connectionless-mode service, unless otherwise explicitly determined, has the following fundamental characteristics:

(a) It requires only a pre-existing association between the peer-entities involved, which determines the characteristics of the data to be transmitted, and a two-party agreement between each peer-entity and the service provider; no dynamic peer-to-peer agreement is involved in an instance of the use of the service.

(b) All of the information required to deliver a unit of data, termed (N)-Unitdata, (destination address, QOS selection, service options, etec.) is presented to the layer providing the connectionless-mode service, together with the user data to be transmitted, in a single network service access which is not required to relate to any other service access.

(c) each unit of data (N-Unitdata) transmitted is entirely self-contained, and can be routed independently by the network service provider.

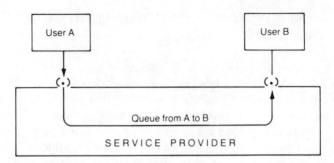

Fig. 4.28 Queue model for connectionless-mode network service.

The model of the connectionless-mode network service (hereafter abbreviated to CLNS) is simple, the principles of the queue model technique used for the CONS can be applied. For CLNS the queue only exists for a single direction, however, as shown in Fig. 4.28 and only one type of object (the Unitdata object) can be placed in the queue.

For what is regarded as the universally guaranteed service, it must be assumed that the network service provider may perform any or all of the following actions:

(a) discard objects which are in the queue;

(b) duplicate objects which are in the queue;

(c) change the order of the objects which are in the queue.

Many practical implementations of connection-mode transmission, however, do exhibit one or more of the following characteristics, for transmissions between a given pair of NSAPs:

(i)   objects will be discarded only after a stated time;

(ii)   objects must be discarded no later than a stated time;

(iii) objects will be discarded only if more than a certain number of objects are in the queue;

(iv)   objects will not be discarded;

(v)   the order of the objects in the queue will not be changed;

(vi)   objects will not be duplicated.

From earlier discussions on the principles of layers and layer services it can be recalled that the protocol within a layer provides the necessary functionality to bridge the gap between the service available from the layer below, and those required by the layer above. In order to design the right

Parameter	N-Facility		N-Report
	Request	Indication	Indication
Destination address		X	X
Service characteristic/ QOS parameter	X	X	X
Reason for report		X	X

Fig. 4.29 Management primitives for the connectionless-mode network service.

kind of protocol, and to minimise duplication of functions in layers it is necessary to be aware of whether only the basic characteristics of connectionless-mode (a) to (c) above or whether characteristics (i) to (iii) above are available.

Service characteristics in categories (i) to (vi) are long-term in nature, sometimes termed static, unlike those QOS characteristics provided on a per-instance basis by the use of parameters set for each invocation of the service (priority for example). These static kinds of characteristics have to be made known to the network service user (Transport entities) prior to any given communication instance or series of instances, so that the appropriate class of Transport Layer protocol may be chosen. ISO 8348 AD1 contains an annex A which provides a model for this situation. The model basically describes the need to convey management and/or control information across the Layer 3/4 boundary. Information is exchanged via the set of primitives shown in Fig. 4.29.

The N-Facility request is issued by the network service user to request information about the characteristics of the service which may be expected to be available for a N-Unitdata request(s) to a given destination NSAP. The corresponding response from the network service provider is contained in the N-Facility indication. The (N)-Report indication is issued by the network service provider, to inform the network service user about a failure to provide the expected QOS or service characteristic. The following service characteristics are defined:

(a) congestion control;

(b) sequence preservation probability;

(c) maximum NSDU lifetime.

If congestion control is available, it would mean that the network service user is willing to be flow-controlled by the network service provider, and therefore the network service user would never be involved with abortive transmissions. This, in turn, of course means that the network service provider is willing to take responsibility for any data already accepted, and not discard it.

Availability of sequence preservation will relieve the network service user from needing a resequencing protocol in the Transport Layer since a simpler retransmission protocol would suffice. The example of use of LLC1 for LAN operation discussed in section 4.4.2.5, is an example of this characteristic.

Knowledge about maximum NSDU lifetime is necessary in order to set the correct retransmission timer values in the Transport protocol class 4. The network service provider will, in general, retain an NSDU for delivery (in queues for delivery) for a period of time equal to the NSDU lifetime parameter.

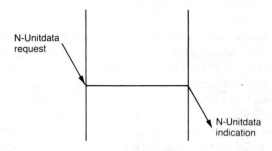

Fig. 4.30 N-Unitdata primitive time sequence diagram.

Parameter	N-Unitdata request	N-Unitdata indication
Source NSAP address	X	X (=)
Destination NSAP address	X	X (=)
Quality of service	X	X
NS-Userdata	X	X (=)

Fig. 4.31 N-Unitdata primitives and parameters.

Dynamic QOS characteristics are invoked per N-Unitdata by parameters associated with the N-Unitdata primitives. The sequence of primitives for a successful communication of a single N-Unitdata is shown in Fig. 4.30. The set of parameters associated with the two primitives are shown in Fig. 4.31. The maximum amount of data that can be sent per N-Unitdata is 64 512 octets. This value has its origin from the particular protocol which has been standardised for provision of the service.

The following dynamic QOS parameters are specified for possible use by the initiators of an N-Unitdata request:

(a) transit delay;

(b) protection;

(c) cost determinants;

(d) residual error probability;

(e) priority;

(f) source routing.

Only the priority parameter is currently defined to be present in the delivered N-Unitdata indication.

The one parameter needing further discussion, since its use is not so obvious, and which is not included in the connection-mode Network service (CONS), is the source routing parameter. This parameter allows the network service user to directly, or indirectly, generate a list of intermediate systems to be included in the path (route) for a given N-Unitdata from source to destination.

Provision is made for two forms of source routing. The first form, referred to as complete source routing, requires that each intermediate-system in the list (and only those intermediate-systems in the list) must be the route taken. If during the course of transfer it is found that the requirements cannot actually be met, then the entire N-Unitdata must be discarded. In this case, the originating user may be informed of the discard by the N-Report indication.

The second form is referred to as partial source routing. In this case, although each intermediate-system in the list must be visited in the order specified, the NSDU may take any other path necessary to arrive at the next intermediate-system in the list. The NSDU shall not be discarded (for source routing related-causes) unless one of the intermediate-systems in the list cannot be reached by any available path.

Source routing is particularly useful in two cases. Complete source routing can be used to provide extra security protection, by excluding 'untrusted' intermediate systems. Complete or partial source routing is useful to assist intermediate systems to reach destination NSAPs in the case where the NSAP address is unknown due to lack of adequate directory information. (Normally an intermediate system would be expected to derive routing information by looking-up the NSAP address in a directory system. Prior to the establishment of such advanced techniques, however, or in the case of addresses not being in the directory, user supplied routing may be required.)

4.5.2. The provision of the connectionless-mode network service (CLNS)

Unlike the CONS, there is only one single protocol defined to provide the CLNS. This protocol is an SNICP designed to operate over any kind of subnetwork, and designed for the internet approach of subnetwork interconnection. The CLNS protocol is defined in ISO 8473. This will be described in some detail since it has not had the exposure of the public network CCITT Recommendations like X.25 or X.21.

The protocol data unit (PDU) consists of the header (PCI) plus network service user data. Since it is possible that the NSDU size is greater than the subnetwork service data unit size, then a segmentation function is required. In the connectionless subnetwork environment it is possible that different subnetworks offer different subnetwork service data unit sizes. The intention is that segmentation should take place as necessary, but that reassembly only be performed at the destination, to avoid complexity and loss of performance at intermediate systems. Reassembly has to involve timing mechanisms, due to the possibility that segments of the entire NSDU may arrive in a random order due to dynamic routing and discard operations within the service provider.

The basic PDU format is shown in Fig. 4.32. Segmentation is the most important function of the protocol and its operation is shown in Fig. 4.33. In this example the original size of the NSDU (i.e. the amount of Transport Layer information) is 510 octets. In this example the underlying subnetwork is assumed to be able to convey and maintain the integrity of a unit up to, say, 650 octets. Thus any initial PDU (i.e. header and data) less than this value can be accommodated. Let us further assume that the length of the header is 128 octets. The figure demonstrates how the various relevant fields are set in the initial PDU, and subsequent derived PDUs. The example shows the case where the initial PDU is segmented into two derived PDUs because, let's say, the next subnetwork component can only convey data units of up to 400 octets.

Note that the length indicator is always the same, because the rules for segmentation require the header to always be the same length.

The Total Length field specifies the entire length of the initial PDU, including both the header and data. This field is not changed in any segment (derived PDU) for the lifetime of the PDU, i.e. 638 for our example.

The values for the fields in the derived PDUs are self-explanatory. The Segment Permitted (SP) flag must always be set to one (i.e. permitted) for ISO purposes. The More Segments (MS) flag indicates whether the data part of a particular derived PDU contains the final data octet of the initial PDU. When it is set to 1 this indicates: (a) that segmentation has taken place; and (b) that the last octet of the original NSDU is not contained in this PDU.

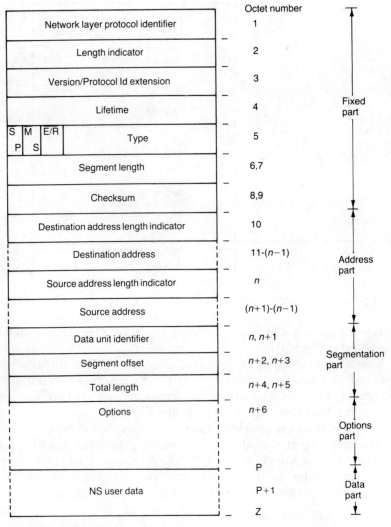

Fig. 4.32 PDU format for connectionless network protocol.

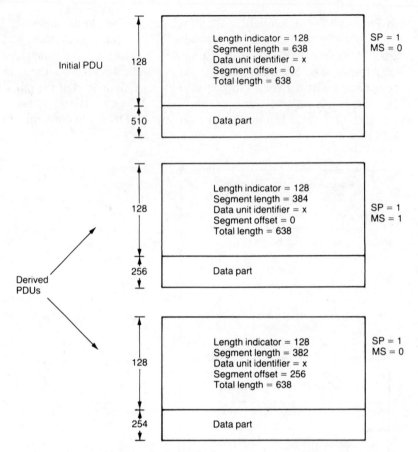

Fig. 4.33 PDU segmentation in the connectionless network protocol.

When it is set to 0 it indicates that the last octet of the original NSDU is contained in this PDU.

The data unit identifier of any and all of derived PDUs is set to the same value as contained in the initial PDU. This ambiguity occurs during the reassembly process (normally at the destination) and is also how error reporting relates error reports to particular PDUs.

The address part is straightforward and carries the full NSAP address.

The options part is used to convey optional parameters, and has the generic encoding shown in Fig. 4.34. The options defined below are not mandatory as far as the initiator is concerned. However, as far as intermediate systems are concerned they fall into three categories:

Type 1   those that must be supported by the service provider, but optionally selected by the initiator;

Type 2   those that, if not supported by an intermediate system, cause the intermediate system to discard the PDU containing these, and to generate an Error Report PDU (ER PDU));

Type 3   those that must be passed transparently across an intermediate system that do not actually support the related service (i.e. not discarded).

Figure 4.35 also summarises the non-parameter driven functions, e.g. segmentation, etc.

The padding parameter exists to permit the initiator of the PDU (i.e. the initial PDU) to build-out the header to a convenient size.

Complete source routing and security are the only two parameters of type 2. Obviously these cannot be passed transparently. If the service required cannot be assured, then the attempt must be abandoned, since to continue could actually generate a breach of security. So for these parameters, if not supported, the PDU is discarded and the reason signalled back to the initiator in an ER PDU. ER PDUs are distinguished from data PDUs in the 5-bit PDU-type field in octet 5 of the header. It should also be noted that the initiator may suppress error reports by use of the E/R (0 = suppress, 1 = do not suppress).

Partial source routing is a type 3 option. The form of address used for both complete and partial source routing is the same as that used for NSAP addresses. Care must be taken to ensure that addresses used for source routing purposes are not allocated also for use as NSAP addresses.

The recording of route parameter enables a list of intermediate systems to be built-up as the PDU finds its way from source to destination. Its use is not quite as simple as that, however, since the size of the PDU header always has to remain the same. Thus enough space must be allocated from the outset.

Bits   8   7   6   5   4   3   2   1

Octets n	Parameter code
n+1	Parameter length (e.g., 'm')
n+2 n+m+1	Parameter value

Fig. 4.34 Generic encoding for parameters of the connectionless network protocol.

Function	Type
PDU composition	1
PDU decomposition	1
Header format analysis	1
PDU lifetime control	1
Route PDU	1
Forward PDU	1
Segment PDU	1
Reassemble PDU	1
Discard PDU	1
Error reporting	1
PDU header error detection	1
Padding	1
Security	2
Complete source routing	2
Partial source routing	3
Priority	3
Record route	3
QOS maintenance	3

Fig. 4.35 Table of the non-parameter driven functions of the connectionless network protocol.

The structure of the parameter fields of source/partial routing and record routing are similar, containing pointers to the active element in the address list. The structure is shown in Fig 4.36 (a) and (b). For recording of route the empty address space, however, is filled-up by adding elements to the front of the list.

The QOS maintenance parameter can be used to indicate relative importance between transit delay/cost, error rate/transit delay, and error rate/cost. Intermediate systems are expected to include this information, with the other relevant information, in decisions about routing (where routing flexibility is available). Priority may be separately specified and also affects routing choices.

The fixed part of the header contains two other fields meriting further discussion, the PDU lifetime field, and the PDU checksum.

The initial value of the lifetime field is set by the originating Network Layer entity in the initial PDU. Every time another Network Layer entity receives and processes a PDU, the value is decremented to reflect the transit delay accumulated in transit from the previous Network Layer entity, plus the time expended in processing the PDU. If the lifetime reaches a value of zero before the PDU is delivered to the destination, the PDU is discarded. The error reporting function is invoked whenever a PDU is discarded under these circumstances. In cases where the initial PDU is segmented into a

Octet

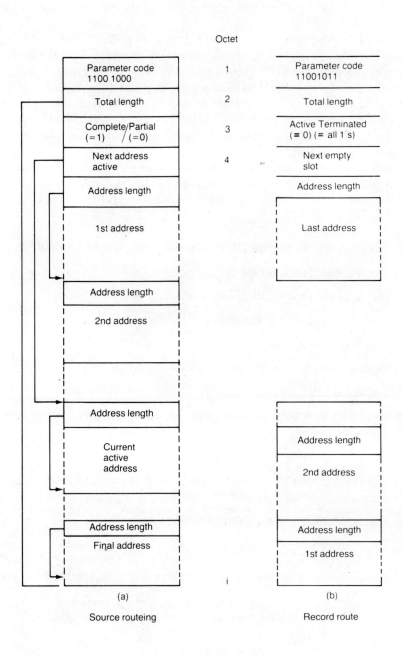

(a)

Source routeing

(b)

Record route

Fig. 4.36 Source/partial routing parameter.

number of derived PDUs, all the derived PDUs have the same value for the lifetime field.

The PDU checksum is initially computed on the initial PDU header. The PDU header fields may be modified by intermediate Network Layer entities. Clearly, the PDU checksum needs to be modified accordingly. However, if the new PDU checksum were computed on the basis of the complete header, any unintentional error occurring in an unmodified part of the header would go undetected. Thus, the checksum is updated only to reflect the modifications made. Any error introduced into an unmodified field can now be detected.

Any time a PDU is discarded, the error-reporting function is invoked. This will result in the generation of an ER PDU, to be sent to the originator of the discarded PDU (provided the E/R flag was set to 1 in the original PDU). The ER PDU:

(a) identifies the discarded PDU (by means of the data unit identifier);

(b) specifies the reason for the discard;

(c) identifies the location at which the discard took place;

(d) contains all or part of the discarded PDU.

A PDU may be discarded for the following reasons:

(a) violation of protocol procedure has occurred;

(b) the checksum is inconsistent with the PDU contents;

(c) the PDU has been received, but due to congestion it cannot be processed;

(d) the header cannot be analysed;

(e) the PDU cannot be segmented and cannot be forwarded because its length exceeds the maximum subnetwork service data unit size;

(f) the destination address is unreachable or unknown;

(g) incorrect or invalid source routing was specified because of a syntax error in the source routing field, an unknown or unreachable address in the source routing field, or a path which is not acceptable for other reasons;

(h) the PDU lifetime has expired or the lifetime expires during reassembly;

(i) the PDU contains an unsupported option(s).

ER PDUs themselves, if discarded, do not result in further ER PDUs being generated.

ISO 8473 contains a formal description of the protocol, based on the Pascal programming language. The protocol is modelled by a finite state automaton governed by a state variable with three values. Its behaviour is defined with respect to individual independent PDUs.

4.6 NETWORK INTERCONNECTION

4.6.1 General principles

Provision of the connection-orientated network service (CONS) is generally specified for the case of a single type subnetwork, e.g. an X.21-based circuit switched subnetwork, LAN of type X, etc.

Having achieved the CONS for a number of different subnetworks, the interconnection of these subnetworks is a straightforward matter and can be achieved by simply mapping one CONS realisation to another. In this case it is immaterial whether the underlying subnetwork is an OSI subnetwork or whether hop-by-hop enhancement is in use, as discussed in section 4.3. It is thus possible to achieve interconnection by using the half-gateway approach in which the interconnection is realised by mapping at the OSI network service level itself (as shown in Fig. 4.37). In practice, such a realisation may be by pure software with a single physical interworking unit or via a hardware realisation of the OSI network service. The routing function can be placed either only in one-half or both. If in both it would be possible to interconnect several 'half-gateway' boxes together to form a network of gateways. As far as can be determined at this time, this method of 'hop-by-hop' enhancement and OSI subnetwork interconnection is favoured as the solution for the connection-orientated mode of operation.

As far as connectionless mode is concerned only one protocol has been defined and this is designed for the 'internet' style of interconnection, for which Fig. 4.8 still applies.

4.6.2 LAN/WAN interconnection issues

The question of interconnection of LANs and WANs such as X.25 subnetworks, merits further attention. We have seen how a connectionless subnetwork may be converted into a connection-orientated network service (i.e. CONS over LLC1) and how, contrariwise, a connection-orientated subnetwork may be converted to provide the connectionless network service. So for LAN/WAN interconnection it is possible to choose either a connection-orientated network service, or a connectionless network service

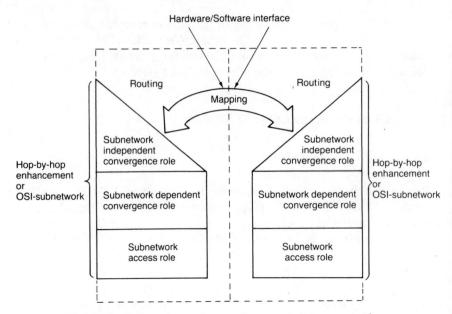

Fig. 4.37 Connection-orientated network interconnection.

for operation over the tandem configuration as shown in fig. 4.38 (a) to (d).

Apart from using personal preference, several factors can be applied to assist the decision. A complete discussion of these issues would require another complete book. A summary of some merits/demerits is outlined below. When considering these issues it should be borne in mind that the basic characteristics of LANs and WANs are almost at opposite ends of a spectrum. LANs have extremely high data rates, low delay, low error rates, short propagation time. Conversely, WANs have low data rates, high error rates (intrinsically) and long propagation delays.

4.6.2.1 Connection-orientated interconnection (COI)

Possible advantages of COI are:

(a) Minimum functionality required in the end systems, i.e. only a simple Transport Layer protocol.

(b) Effective congestion control and fairness. Flow control can be applied for every gateway and intergateway link, permitting back-pressure to be exerted on a per connection basis.

(c) Hop-by-hop enhancement can be applied so that any deficiencies can be remedied at their source (including retransmissions). The 'internet'

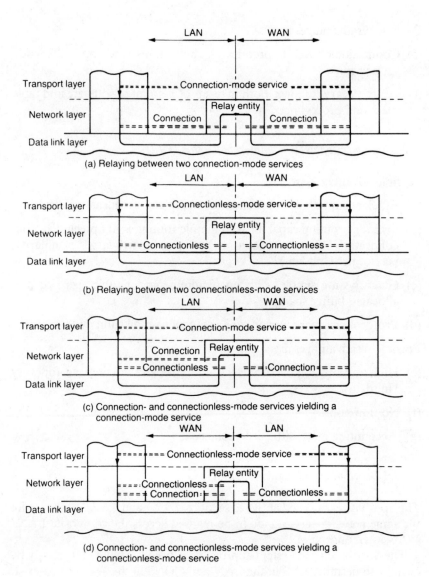

Fig. 4.38 Permitted relaying configurations.

approach for CLNS can only achieve enhancement on an end-to-end basis, primarily at the Transport Layer which is inefficient and costly.

(d) Accounting and management information can be achieved on a per-connection basis.

Possible disadvantages are:

(e) Connectionless and broadcast applications are not efficiently supported.

(f) Gateways are complex, both in terms of amount of software and table space and processing.

4.6.2.2 Connectionless-mode interconnection (CLI)

Possible advantages are:

(a) Efficient support of connectionless-mode and broadcast applications:

(b) Dynamic routing capabilities (dynamic routing is not precluded from connection-orientated operation, but operates on a larger granularity, i.e. not per Unitdata);

(c) Gateways are simple, with no requirements for table space or pre-allocated buffer space;

(d) Protocol variety reduction, one protocol over all subnetworks.

Possible disadvantages are:

(e) High transmission overheads (i.e. addresses and activities etc. for every Unitdata).

(f) No flow control.

(g) Accounting more difficult to achieve.

4.6.2.3 Distributed end-system approach

The approaches outlined above in sections 4.6.2.1 and 4.6.2.2 both require the same network service type to be realised across the composite LAN/ WAN interconnection.

The distributed end-system approach has been formulated primarily by ECMA to permit CLNS in the LAN and CONS in the WAN to co-exist. This is achieved by joining the two environments together at the Transport Layer. This is based on the assumption that the applications are primarily concerned with the nature of the Transport service and not how it is realised. The means of interconnection in this case adopts the half-gateway principle described previously but for this case the mapping is achieved at the level of the Transport Layer service rather than the Network service. This effectively produces a relay at the Transport Layer, which is expressly forbidden in the OSI architecture. However, it is a popular method of

LAN/WAN interconnection and can be said to conform only by considering the whole of the LAN environment as an OSI end-system, whose internal structure is not visible externally. This approach is shown in Fig. 4.39.

Apart from being able to reconcile differences of Network service type, this solution has an advantage of enabling different classes of Transport protocol to exist for a given Transport connection, with conversion taking place between one class and another in the Transport Layer relay.

4.7 NETWORK LAYER ADDRESSING

4.7.1 General principles

ISO has produced ISO 8348: AD2 specifying a global addressing scheme for the identification of NSAPs. The objective is that one single address should be all that is necessary to reach a given destination, irrespective of the number of subnetworks between source and destination and irrespective of the number of subnetworks to which the destination is attached and irrespective of the location of the caller. In short, from anywhere to anywhere with a global identifier.

CCITT produces Recommendations for the numbering plans (address

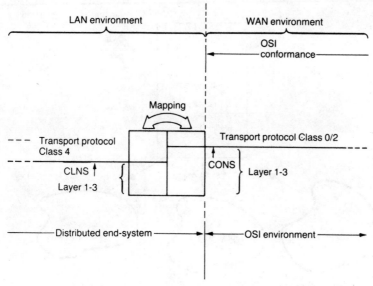

Fig. 4.39 Distributed end system approach.

schemes) for public networks, e.g. F.69 for Telex, X.121 for public data networks (PDNs), E.163 for the Public Switched Telephone Network (PSTN), etc. These numbering plans are different for each network and generally only pinpoint and identify the point of attachment to the particular public network, i.e. in OSI terms, the subnetwork. Such addresses are called subnetwork point-of-attachment (SNPA) addresses. The objective stated above can only, therefore, be met by creating a logical distinction between an NSAP address and the SNPA address to accommodate the OSI end-system connected to a number of different subnetworks, as shown in Fig. 4.40. There may be a one-to-one, many-to-one or many-to-many relationship between SNPAs and NSAPs.

Prior to the OSI initiative, no schemes were available for the global identification of private networks *per se*. Thus, the multiple interconnection of private networks (even those belonging to the same organisation) was difficult. The interconnection of a private network to a public network was equally difficult unless the private network was extremely small and able to be regarded as a simple set of extensions to the public network.

Finally, there is the case where there are several subnetworks in tandem. Again, the objective of using a single address is paramount. The alternative of the source specifying a list of all the intermediate subnetwork addresses is not acceptable because:

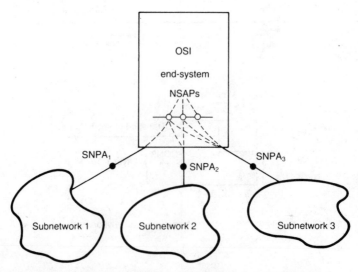

Fig. 4.40 Distinction between SNPAs and NSAPs.

(a) the number of addresses to be supplied can increase beyond reasonable bounds;

(b) changes in subnetwork configurations causes changes in requirements placed on the caller;

(c) the caller has to be aware of routing possibilities;

(d) it precludes route optimisation at intermediate system points.

All the above considerations lead to the global Network Layer addressing standard ISO 8348: AD2 which can be used to allocate NSAP addresses which are globally unique and logically independent from any subnetwork addressing scheme.

The operation of such a scheme is shown in Fig. 4.41 (a) and (b). A domain may be regarded as a network administration authority (e.g. national, private, etc.). Suppose we need to contact F. Bloggs. Use of a directory (manually or electronically by various means) will produce the global address of Y1234, meaning that F. Bloggs is globally known by specifying 1234 together with the domain identifer Y. F. Bloggs can be reached from callers in other domains X or Z, via appropriate routes (if they exist) involving subnetworks Q, P and R. Now the actual route chosen will depend on the number of routes that actually exist and how the route is chosen from the set of possible routes. The latter choice depends on what optimisation or other factors are taken into account when the choice is made, e.g. balancing loads, tariff aspects, quality of service, etc.

The calling system X1234 puts the destination NSAP address Y1234 in the N-Connect Request. One other piece of information is required, i.e. the address of the intermediate system Il which is the gateway from SNX to the outside world. There are two ways in which this could be obtained, either from the caller or by mechanisms within SNX, if SNX itself understands OSI global NSAP addresses. This will depend on the commercial decisions about facilities offered by SNX.

When the intermediate system Il receives the global NSAP address it will use its directory information to choose the next subnetwork to be traversed. In Fig. 4.41 it is assumed that this results in the subnetwork address SNR 1357, which is the address of the next required intermediate system I2 capable of interconnecting SNR to SNY. The global NSAP address is then relayed to I2. I2 applies the same process. However, in this case it so happens that SNY has been already reached and thus Y1234 may be directly usable as the subnetwork address. This is not necessarily the case, however, and a further translation to a subnetwork-dependent address may be required.

It has to be said that this scheme is not trivial in view of the translation tables required in gateways and the need for associated directory systems.

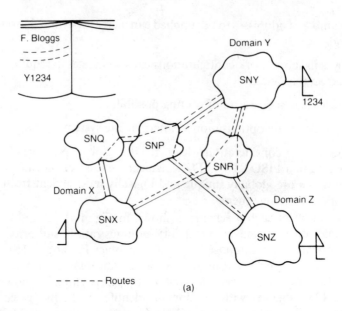

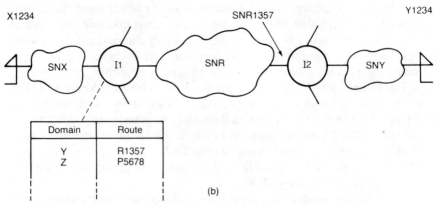

Fig. 4.41 Global addressing principles.

However, it is the only way to achieve the long-term objective and the potential flexibility associated with such a scheme. Mobile addressing, for example, could be accommodated as well as sophisticated call redirection, generic addressing, logical addressing schemes.

It must also be possible to utilise existing networks without changing their individual addressing schemes or internal routing implementations, even where they are being used as part of a composite OSI Network connection between other networks. An OSI Network connection can span a number of real, separate networks which are interconnected by gateways (private or public) in such a way as to provide the appearance to the end-users of a single composite connection. In some cases the individual networks involved in such a chain (subnetworks within OSI) may not know that they are being used in such a fashion. Indeed, for generality and flexibility it is essential that individual subnetworks should not need to know how they are being used.

As far as Network Layer protocols are concerned, these considerations lead to the requirement for the capability with the Network Layer to convey two address fields across real subnetworks; one for the individual subnetwork address scheme, and the other for the global OSI Network Layer address.

It has already been seen that X.25 (1984) has this capability and that X.25 (1980) requires the additional SNDCP to provide this capability.

An individual subnetwork which inherently supports the OSI Network service (X.25 (1984), for example) may participate in the OSI scheme in addition to the X.121 scheme. Thus, two schemes could co-exist on such subnetworks based on commercial decisions. Furthermore, participation in the OSI-based scheme could be implemented either within the subnetwork proper or as a value-added service via independent gateways.

4.7.2 The global address scheme

A global scheme needs to be controlled to ensure that it is globally agreed and so that addresses can be unambiguously allocated. Furthermore, the scheme must not be too rigid. Provision for different kinds of address schemes must be made, e.g. for geographic and non-geographic schemes, existing worldwide schemes and national requirements. This flexibility, in turn, demands a globally understood method of distinguishing between different forms of address schemes.

The above considerations naturally lead to a tree-like structure to permit devolution of authority at certain points in the address structure. Many existing address schemes already have simple versions of such a structure. Take, for example, a UK telephone number 44 1251 2190 extension 356. The 44 is agreed by all countries to identify the UK and thus 44 belongs to a certain international domain of allocation. The next digit 1 identifies London and thus this digit is operating at the level of a national regional domain. The remaining digits belong to further subdomains within London and finally the extension numbers are allocated by the telephone user.

Domains are characterised by the authority that administers the domain and by the rules that have been established by that authority for specifying identifiers and identifying subdomains.

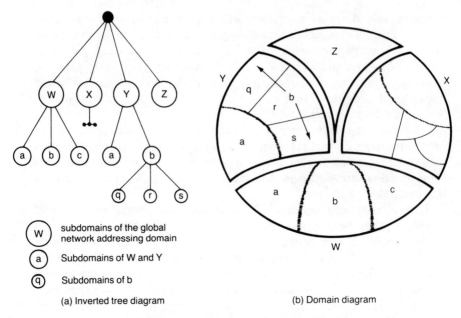

(a) Inverted tree diagram

(b) Domain diagram

Fig. 4.42 Address hierarchies and domains.

Equivalent graphical representations of these concepts by an inverted tree diagram and a domain diagram are shown in Fig. 4.42 (a) and (b).

The operation of an authority is independent from that of any other authority at the same level in the hierarchy and is only subject to the rules imposed by the parent authority.

4.7.3 The structure of OSI Network addresses

The number of domains possible and the lengths of addresses are inextricably bound together. For example, suppose there was only one domain and this domain authority could allocate a unique number to every NSAP in the world. In this case, not very many digits would actually be required. Such an address would have no structure, the relationship between an NSAP and its geographic location, or type of network, would be completely arbitrary. Implementation of such a scheme would be extremely complex and inefficient, involving access to massive directories to ascertain required routing information. Whilst directory systems are desirable for flexibility, a single directory with worldwide scope is not practical and some structured division of the problem into manageable parts is required.

So the length of the address must be sufficient to permit a reasonable number of domains. The eventual decision on the maximum length was thus a compromise based on a reasonable number of domains, within a reasonably manageable total address length. The maximum length of the address is constrained by two factors: a maximum length of 20 octets when encoded in binary form, or 40 decimal digits when encoded in decimal digit form.

Thus, a Network Layer protocol capable of conveying either a maximum of 20 binary octets or 40 decimal digits is capable of encoding the full semantic content of any OSI NSAP address. It should be noted that $10^{40}-1$ is less than $2^{160}-1$, thus the maximum potential of 20 binary octets can never be realised. A given Network Layer protocol determines the concrete syntax *a priori* (e.g. if X.25 is used, the concrete syntax is always in terms of decimal digits, one digit per semi-octet).

The basic structure of the Network Layer address is shown in Fig. 4.43.

The AFI consists of an integer in the range 0 to 99 with an abstract syntax of two decimal digits. Different values of the AFI specify the format of the IDI, the authority responsible for allocating values of the IDI, the type of abstract syntax of the DSP. The abstract syntax of the IDI is always decimal digits, whereas the DSP can have one of four abstract syntaxes, binary, decimal, international character (ISO 646) or national character.

Values of the AFI in abstract syntax are shown in Fig. 4.44. Seven different IDIs are identified with binary and decimal DSPs together with character syntaxes for local usage. AFI values currently standardised are in the range 36 to 51. Other blocks of AFI values are reserved for either ISO or CCITT or joint ISO/CCITT assignment. Values 00 to 09 are permanently reserved and 0 is used as an escape for a special case in which the

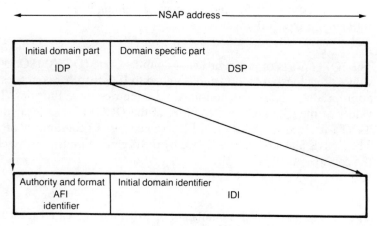

Fig. 4.43 NSAP address structure.

DSP syntax / IDI syntax	Decimal	Binary	Character (ISO 646)	National character
X.121	36	37		
ISO DCC	38	39		
F.69	40	41		
E.163	42	43		
E.164	44	45		
ISO ICD	46	47		
Local	48	49	50	51

Fig. 4.44 Assigned AFI values.

address field only contains the latter part of the total NSAP address and where the first part is carried in a subnetwork address field.

Four of the address schemes are based on CCITT-defined numbering plans, i.e. X.121 for public data networks, F.69 for Telex, E.163 for Public Switched Telephone Network and E.164 for the Integrated Services Digital Network. In these cases, the DSP constitutes an extension address to the public network address.

The IDI formats are defined as follows:

(a) *X.121*

The IDI consists of a sequence of up to 14 digits allocated according to CCITT Recommendation X121. The X.121 number identifies an authority responsible for allocating and assigning values of the DSP. IDP length: Up to 16 digits.

(b) *ISO DCC*

The IDI consists of a 3-digit Data Country Code (DCC). ISO DCC values are allocated by ISO and assigned to ISO-member countries or appropriately sponsored non-member countries or authorities. The values of the ISO DCC are a subset of the DCC values allocated by CCITT in Recommendation X.121 to countries or geographical areas. The DSP is allocated and assigned by the organisation that represents the country identified by the DCC. IDP length: 5 digits.

(c) *F.69*

The IDI consists of a telex number of up to 8 digits, allocated according to CCITT Recommendation F.69, commencing with a 2- or 3-digit destination code. The telex number identifies an authority responsible for allocating and assigning values of the DSP.

IDP length: Up to 10 digits.

(d) *E.163*

The IDI consists of a Public Switched Telephone Network (PSTN) number of up to 12 digits allocated according to CCITT Recommendation E.163, commencing with the PSTN country code. The PSTN number identifies an authority responsible for allocating and assigning values of the DSP.

IDP length: Up to 14 digits.

(e) *E.164*

The IDI consists of an ISDN number of up to 15 digits allocated according to CCITT Recommendation E.164, commencing with the ISDN country code. The ISDN number identifies an authority responsible for allocating and assigning values of the DSP.

IDP length: Up to 17 digits.

(f) *ISO 6523-ICD*

The IDI consists of a 4-digit International Code Designator (ICD) allocated according to ISO 6523. The ICD identifies an organisational authority responsible for allocating and assigning values of the DSP.

IDP length: Up to 6 digits.

(g) *Local*

The IDI is null.

IDP length: 2 digits.

The actual carriage of NSAP addresses in Network Layer protocols is relatively complex, because of possible encoding restrictions on particular address fields and the combination of address fields to be used.

The recognition of the encoding restrictions has led to the development of what are termed two concrete syntaxes, one in binary and the other in decimal. The first is for use in protocols which do not place any encoding restrictions. Rules are specified for representing decimal, binary and character-based abstract syntaxes into binary codes. The second is for use in protocols which only permit decimal digits to be transferred.

For the binary concrete syntax it is assumed that the protocol is capable of conveying unrestricted bit patterns. Thus, the binary syntax specifies how decimal digits and character strings should be encoded. Decimal digits are encoded using binary coded decimal codes with a 4-bit semi-octet. Characters are encoded by taking the numeric value of the character and coding it as two decimal digits in the range 0 to 95 (only characters 32 to 127 may be used and 32 is subtracted to achieve a 2-digit representation).

For the decimal concrete syntax the AFI and IDI values are represented by decimal digits. A decimal representation of a binary string is achieved by taking each pair of octets and converting this into two 3-digit decimal

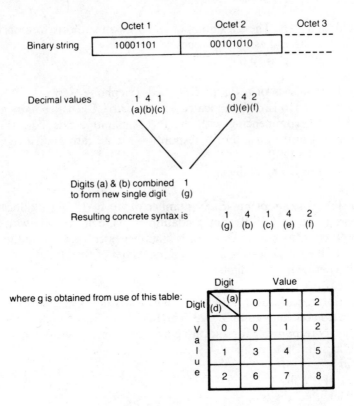

Fig. 4.45 Conversion from binary to decimal concrete syntax.

numbers, each in the range 0 to 255. Since the most significant digit of each of these can only take values 0, 1 or 2, it is possible to combine these two into a single digit in the range 0 to 8. The process is shown in Fig. 4.45. The last octet of an odd-number octet string is simply converted into a 3-digit decimal number 0 to 255. For the decimal concrete syntax the actual protocol encoding may vary depending on how the particular protocol carries decimal digits.

As far as use of various protocol address fields is concerned, a number of cases need to be considered. X.25 (1984) is chosen for the purpose of the examples because it includes its own subnetwork address field (capable of carrying a limited OSI NSAP address) together with a full OSI NSAP address field (the CCITT-specified DTE facility – calling/called address extension). The examples are equally applicable to X.25 (1980) and the X.25 (1980) SNDCP and other combinations of protocols offering a subnetwork-dependent address field together with an OSI NSAP address field.

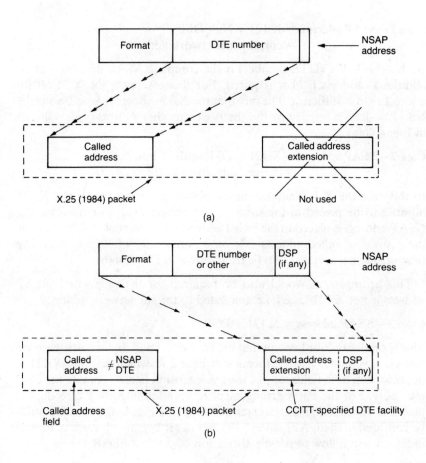

(a)

(b)

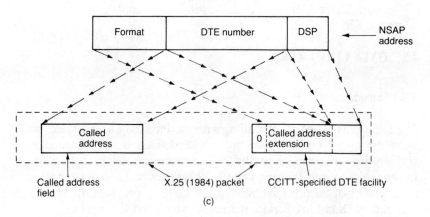

(c)

Fig. 4.46 Carrying the NSAP address in SNACPs.

Case 1 – NSAP address = X.121 + Null DSP
= current subnetwork address

In this the X.121 (DTE number) is the complete NSAP address and no additional address field is required. For this case even the X.25 (1980) address field is sufficient. The fact that the X.25 subnetwork address is the NSAP address is implicit by the absence of another address field as shown in Fig. 4.46(a).

Case 2 – NSAP address = X.121 + DSP (null or non-null)
= current subnetwork address

In this case the X.121 number in the NSAP is different from the X.121 number to be placed in the subnetwork address field. For this case the NSAP address is placed in the called address extension field (X.25 1984) or the equivalent called address extension parcumeter of X.25 (1980). The relationship between the two fields is shown in Fig. 4.46(b).

This arrangement would also be required for the case where NSAP address is not X.121-based, i.e. unrelated to the subnetwork address.

Case 3 – NSAP address = X.121 + DSP

This is similar to case 1 except that the DSP is not null. In this case there are two choices: either use the formats of case 2 in which case the X.121 is duplicated in both fields, or just use the extension field to contain the DSP part only. For the latter arrangement to avoid ambiguity a zero digit is placed as the first digit of the extension field (see Fig. 4.46(c)). This cannot be confused with an AFI since AFIs can never begin with zero. A format indicator will follow to permit distinction of syntax of DSP.

Network Layer addressing is of vital importance to OSI. The relationship of Network Layer addressing to addressing in the upper layers is discussed in section 3.9.

4.8   DATA LINK LAYER

4.8.1 Introduction

The Data Link Layer service, along with the Physical Layer service, differs from the higher layer in that the service defined is not intended to be globally available throughout the OSI environment. Relays in the Network Layer can decouple differing Data Link services and it is not, therefore, a requirement of an open system that its Layer 2 protocol meets a single mandatory Data Link Service definition which is defined for OSI.

Thus, the service definition for this Layer has a different objective to those in the other layers which are primarily aimed at aiding the development of one or more protocol standards, conformance to one of which will be required for an open system. The service definition for this layer is aimed at helping network designers in the development of a new protocol by providing an outline of functional requirements.

The service definition, ISO 8886, also acts as a basis for a specific service definition. From a comparison of the functionality of a Data Link protocol to the OSI Data Link service definition, the Data Link service provided by that protocol should be easily derivable by additions and deletions to the OSI Data Link service. This specific Data Link service can be used to decide on the extra functionality required of the Network protocol in order to provide the OSI Network service (which is a global and mandatory requirement).

Thus each Layer 2 protocol (e.g. ISO 7776 (LAPB), ISO 8802 (LLC), and CCITT Q.921 (LAPD)) would have a specific Data Link service definition associated with it.

When considering LANs, the medium access standards (ISO 8802.3, ISO 8802.4, ISO 8802.5 and ISO 8802.6) are not applicable, since the functionality presented to the Network Layer is determined by the type of Logic Link Control (LLC) implemented over the medium access standard. The medium access technique used only influences certain QOS characteristics.

4.8.2 General requirements

As with terminals, the protocols implemented in the lower layers of an open system are determined in most cases by the network to which the system is connected. The major exceptions to this are the data transfer protocols used on circuit-switched connections. In these latter cases, the Data Link protocol can be subject to bilateral agreement or, sometimes, protocol negotiation.

This means that a wide variety of protocols will be implemented in the Data Link Layer so that systems can be connected to a wide variety of network technologies. Therefore, an equally wide variety of Data Link services will be needed to describe the functions made available by each of the protocols.

It is not possible to achieve service levelling (i.e. a single common service definition) at this layer since many networks define a Layer 3 access protocol which uses the full functionality of the associated Layer 2 access protocol. Any attempt at levelling the Layer 2 services would therefore affect the Sub-network Access protocols and, hence, the network operation.

If OSI is to be implemented at all it must make maximum use of existing networks, both public and private, so service levelling is not achieved at Layer 2.

The availability of parameters at this service boundary will also give an indication of the physical configuration used by the network. Some Data Link protocols will be operating over a point-to-point physical circuit where addressing is redundant (e.g. LAPB of X.25). Others may be operating over a multi-end-point physical configuration where the destination LSAP must be identified from all possible LSAPs available (e.g. for LAN or multi-drop operation).

4.8.2.1 Requirements for connection establishment

Most protocols existing at the Data Link Layer have a range of options which will affect the service offered to the Network Layer in terms of its quality rather than its function specification. One example of this is the multilink procedure described in LAPB of X.25 which would add to the resilience, (i.e. ability to operate when one or more links have failed) and throughput of the data link connection compared to the single link operation. In most cases, options within each protocol are selected by their availability in the network access protocols or chosen at the time of attachment of the open system to the network (i.e. contracted with the network administration). In these cases the quality of the Data Link service is fixed and the parameters are not available for selection or negotiation at the service boundary although the quality will be known to the higher layer. However, in some cases it will be possible to negotiate the use or non-use of certain protocol options on a Data Link connection dynamically. This is particularly true where the connection is being established between systems using a switched circuit so that there is no *a priori* knowledge of the protocol options available (e.g. 32-bit FCS in HDLC).

The quality of service (QOS) available to a user can vary considerably in cases where several connections are competing for the use of a limited shared resource, e.g. the physical medium in a LAN environment. Protocols which operate in such environments often have priority parameters which can be set to ensure that the higher priority connections have the best chance of gaining control of the physical medium in a contention situation. The manipulation of such a parameter would obviously affect the quality of the Data Link service observed by the user.

Where QOS parameters are used to select protocol options (e.g. use of checksum type) then negotiation of those protocol options takes place during connection establishment between the Data Link Layers of the systems involved. The initiating user is then informed by the Data Link provider of the QOS achieved. The accepting user may also be informed by the Data Link service provider of the QOS achieved or the user may extract the information from the Network protocol.

Where parameters are explicitly conveyed by the Data Link protocol both users are informed of the value of this parameter by the Data Link service provider.

The 'use of expedited' parameter in Connection Request primitives allows systems using Data Link services which provide Expedited data to negotiate the use of this service with the peer Data Link service user.

Some services may allow the transfer of a small number of octets of user data with the Connect Request protocol. This data field can be used to convey Network Layer information which the receiving user may take into consideration when deciding whether to accept or reject the call attempt.

The address parameters are used to convey information about the service user to which the connection is to be established or from which the connection request has come. These parameters are included in the service definition to provide functional description for the protocols. Protocols which operate between only two systems and only use an 'address field' in the protocol to differentiate between frame types (e.g. LAPB) would not have an address parameter in the *service* since no addressing function is present as far as the user is concerned.

In fact at the Data Link Layer the addressing function is really a destination selection generally achieved by broadcasting the message to all the devices connected to the physical medium and ensuring that all the devices except the one with the correct address ignores it. The presence of an address parameter does not imply routing since there is no choice of alternatives when travelling from one station to another.

Summarising these requirements, it can be seen that the Data Link service will contain many parameters which are provider options and specific Data Link service definitions would vary widely in terms of the parameters used. It should be noted that the parameters, where used, are likely to have a different effect on the operation of the protocol to the parameters with similar names in higher layer service definitions.

4.8.2.2 Connection Release service

This service is needed to perform two distinct functions:

(a) release of an established connection, by either the user or provider of the Data Link service;

(b) the refusal of a connection attempt, by either the user or provider of the Data Link service.

When the Connection Refusal service is used there may be some information available at one end in terms of the reason or source of the release which is needed by the service user to determine future action. For example, if a connection attempt is unsuccessful due to the initiating end receiving a protocol refusal, as opposed to no reply at all thereby causing a timeout, then this could affect the timing of a retry by the higher layer.

However, none of the existing protocols provide for the transfer of such information and it is unlikely that such parameters will be generally available for the release of established connections in the near future.

Although all the available protocols allow for a confirmation of the release of a disconnection, none of them allow release request to be refused, nor does the release involve a negotiation of any kind. The Release service is unconfirmed since the user need only know that a release has been initiated, either because he sent the request or has received an indication, in order to know that the connection is no longer available. A local indication between the layers of a system indicating the release of resources, e.g. buffers, may be useful but this would not be visible externally and hence is not modelled in the service definition.

4.8.2.3 Data Transfer service

There are two basic types of data transfer which may be required by the Data Link service user. One is data transfer outside the context of a connection (i.e. connectionless) and the other is data transfer within the context of a connection (i.e. connection-orientated).

In connection-orientated data transfer the units of data being transferred are related in sequence and limited in length. This limitation length is usually known at the time of implementation and thus the subject of *a priori* knowledge. In the case of a switched circuit physical connection, where such knowledge may not be available, then the XID frame within HDLC may be used to establish this parameter.

Within a connection, where flow control may be exerted on the data, there are two distinct types of Data Transfer service which may be available. Normal data transfer is always available and subject to the flow control. Expedited data transfer is used to describe the service supplied to the users where data is transferred when it is not subject to flow control. This service is only available if the negotiation for the use of Expedited data has been successfully completed during connection establishment.

For connectionless data transfer, the unit of information being transfered by the Data Link service is independent of information transferred in earlier or later invocations of the service. Also, these transfers are not preceded by a connection establishment phase and, hence, no parameter negotiation has been done, nor has any association with the destination been established. Therefore, the Unitdata primitive needs several parameters associated with it which pass information not required for connection-orientated data transfer (because in the connection-orientated case, the information was transferred during the establishment phase).

The parameters that may be present in addition to the user data parameter are source and destination addresses, and priority for access to shared media. In general, the other QOS parameters which were used in connection establishment for protocol option selection are not used here because the protocol subsets would have been predetermined and changing them for every data transfer is not practicable. As with connection-orientated data transfer the size of the data which may be transferred in a single invocation is limited.

4.8.2.4 Reset service

This service is to fulfil a user requirement to resynchronise both ends of a Data Link connection without releasing the connection. Any data which has been sent but not delivered or delivered but not acknowledged will remain unacknowledged and any flow control or error conditions existing in the underlying protocol will be cleared. Since the connection is cleared of all data and all acknowledgements, the higher-layer protocol is relied upon to recover those data losses. This service is confirmed and thus makes use of all four primitives.

The provider may initiate this service so that the protocol machine in either the local or remote device can indicate that it has detected an error from which it is unable to recover. This service may not be available in all Data Link services and it is unlikely, given the currently available protocols, that any additional information in the form of cause codes, will be available. As with the Connection Release service, some information of this type may be available at the local device and thus made available to one user as local information.

4.8.3 Specific Data Link service definitions

Having studied the principles of a general Data Link service definition the following are examples of specific Data Link services for some common Data Link protocols.

4.8.3.1 HDLC LAPB (X.25 frame without multilink)

The set of primitives and associated parameters used to describe the service provided by this protocol is given in Fig. 4.47.

Phase	Service	Primitive	Parameter
Connection establishment	Connection establishment	DL-Connect request	None
		DL-Connect indication	None
		DL-Connect response	None
		DL-Connect confirm	None
Data Transfer	Data Transfer	DL-Data Request	DL-user data
		DL-Data Indication	DL-user data
	Reset	DL-Reset Request	None
		DL-Reset Indication	Cause
		DL-Reset Response	None
		DL-Reset Confirm	None
Connection Release	Connection Release	Disconnect request	None
		Disconnect indication	Cause

Fig. 4.47 Table of data link service primitives for LAPB.

The lack of parameters in this service definition reflects the fact that all the options of protocol operation will be fixed at the time of terminal registration. The connection request generates a SABM which upon receipt causes the receiving terminal to pass up a Connect Indication primitive. The Connection Response and Confirm primitives are similarly related by UA frames. The same protocol exchange may be used when a connection is already established and this case is modelled using the equivalent Reset primitives. Release is achieved using the DISC frame and the associated UA response.

Data transfer is modelled using only two primitives, DL-Data Request and DL-Data Indication (see Fig. 4.47). The flow control inherent in the protocol is not explicitly shown in the service boundary since it is a purely local service, but some method of flow-controlling the primitives transferred across the boundary will be required. The exception conditions which the protocol machine may enter are not reflected in the service boundary and would only be visible to the higher layer as flow control. The result of such an exception condition may be a Reset or Connection Release which is indicated across the service boundary as a 'provider-initiated' service invocation.

When the exception condition was caused by a Frame Reject response frame then the receiving LAPB must either Reset or Disconnect the connection. In such a situation a cause parameter may be associated with the local indication primitive and used to pass the additional reasons for the provider action to the Network Layer. This is the only situation where these optional parameters are used in this service definition.

A collection of time sequence diagrams relating the primitive actions to the associated LAPB protocol actions are given in Fig. 4.48 to 4.53. Figure 4.53 shows a sequence of information transfers and the use of protocol actions by the layer 2 which do not always result in service primitives being issued. As mentioned earlier, if the recovery shown in Fig. 4.53 had failed, then the protocol would have generated a Reset which can be modelled using either of the diagrams in Fig. 4.52 (a) or (b).

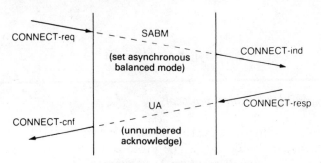

Fig. 4.48 DL connect accepted.

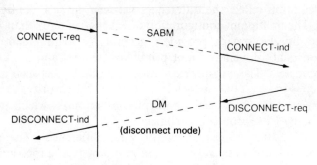

Fig. 4.49 DL connect refused.

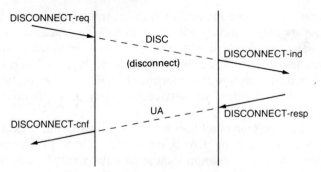

Fig. 4.50 DL connect release.

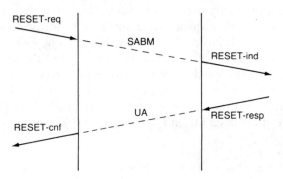

Fig. 4.51 DL reset.

4.8.3.2 Logical link control protocol

The primitive set for the logical link control developed by IEEE for use over LANs is shown in Fig. 4.54. The protocol described is class II which incorporates both type 1 and type 2 operation.

The differences between this service definition and that of the LAPB service definition reflect the differences between physical configurations involved. The multipoint configuration of LANS (i.e. more than 2 LSAPs supported by the service provider) generates a requirement for the connection establishment attempt primitives (request and indication) to carry parameters. These parameters allow the LSAPs (between which the connection is to be established) to be identified. The Priority parameter is used to set the priority for access to the physical resources which are shared between many LLC protocol machines.

The LLC protocols also allow for some differences. The protocol implements an unnumbered information frame-type procedure which allows the service to provide a connectionless data transfer modelled with

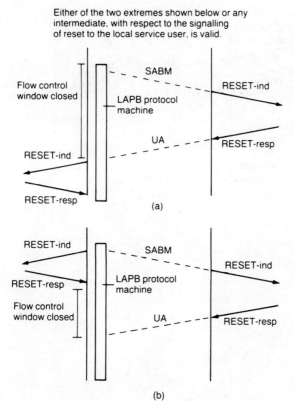

Either of the two extremes shown below or any
intermediate, with respect to the signalling
of reset to the local service user, is valid.

Fig. 4.52 DL service provider initiated reset.

Unitdata primitives. In the LLC definition the Reset and Disconnect Indication primitives also have an associated cause parameter. However, here the parameter is no longer optional and has a value which indicates that the service (Reset or Disconnection) was caused by protocol action (i.e. originated from the other station on the LAN) or by the local LLC protocol machine. If the latter is the case then the value of this parameter indicates the cause which generated service. The values which indicate 'remote protocol action' as the cause of the reset or disconnect must be interpreted as 'cause unknown' since it is not possible for the local LLC to know if the protocol was generated as a result of a remote LLC user request or an autonomous decision by the remote LLC provider. This is as a result of the independence of OSI layers which means that a service user has no knowledge of the structure of the layer providing the service and in particular the DL service user is not aware of the division between protocol machines providing the DL service.

(the numbers in [ ] are sequence numbers)

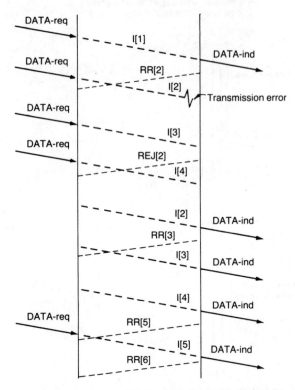

Fig. 4.53 DL protocol recovery from an error.

The differences between the above service definition and that given in the LLC documents stem from a difference in modelling technique. The LLC service definition described in ISO 8802.2 is based upon a three-primitive model of confirmed services as shown in Fig. 4.55. In the ISO Data Link service definition, which is aligned with all other layer service definitions for OSI, a four-primitive model is used as shown in Fig. 4.56. In order to understand the differences between these service definitions let us consider the connection establishment mechanism as modelled using each method.

Both represent the start of a connection attempt by one of the service users issuing a Connect Request primitive which results in the transmission of a set mode protocol command.

Upon receiving this command the LLC description has the LLC protocol machine make the decision to accept or reject the connection and send a Connection Indication primitive to the local user to indicate the decision. The DLS description shows the LLC layer issuing a Connection Indication primitive immediately. The user then decides to accept (and issues a connect

Phase	Service	Primitives	Parameters
Connection establishment	Connection establishment	DL-Connect Request	Calling Address Called Address QOS (priority)
		DL-Connect Indication	Calling Address Called Address QOS (priority)
		DL-Connect Response	QOS (priority)
		DL-Connect Confirm	QOS (priority)
Data Transfer	Data Transfer	DL-Data Request	DL-User Data
		DL-Data Indication	DL-User Data
	Reset	DL-Reset Request	None
		DL-Reset Indication	Cause
		DL-Reset Response	None
		DL-Reset Confirm	None
Connection Release	Connection Release	DL-Disconnect Request	None
		DL-Disconnect Indication	Cause
Any	Connectionless Data Transfer	DL-Unit Data Request	Calling Address Called Address QOS (priority) DL-User Data
		DL-Unit Data Indication	Calling Address Called Address QOS (priority) DL-User Data

Fig. 4.54 Table of data link service primitives for LLC.

response primitive to the LLC layer) or reject (and issues a disconnect request primitive to the LLC layer) the connection attempt. Both models show the LLC issuing a UA response if the connection is accepted or a DM response if the connection is rejected.

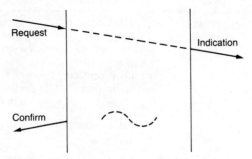

Fig. 4.55 Three-primitive sequence.

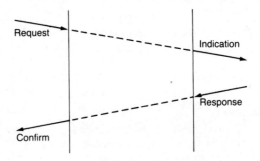

Fig. 4.56 Four-primitive sequence.

Upon receipt of the response, the LLC description shows a Connect Confirm primitive being issued to the user indicating acceptance or rejection of the connection attempt. The DLS description shows a Connect Confirm primitive being issued to indicate acceptance and a Disconnect Indication primitive being used to indicate rejection of the connection attempt.

Notice that in the three-primitive description the Indication and Confirm primitives indicate a successful or failed connection attempt whereas in the four-primitive DLS description the Indication primitive precedes the final decision and the Confirm primitive always indicates a success.

Thus both models show the same possible protocol exchanges and both service users are informed of a successful or failed connection attempt. The apparent difference as to where the decision is made is less clear, considering that some of the information which may be used to decide to reject the call in LLC properly belongs to the Network Layer. In other words, the LLC machine has previously been informed of the conditions under which the connection is to be rejected. This is equivalent to the user response to a Connect indication in the DLS description.

The LLC description more closely models the actual interface which could be seen if the LAN access protocol machine was implemented separately from the LAN user. This is more obvious when considering flow control. The LLC model uses explicit primitives for this function which is local to the system and, hence, cannot be explicitly modelled by the DLS description.

4.9 PHYSICAL LAYER

4.9.1 Introduction

The objective of the Physical service is to describe in abstract terms the characteristics of the media being used to interconnect systems of which the higher layers of the system need to be aware. Thus an important objective of the Physical Layer is to mask, as far as is possible, the characteristics of physical media in order to maximise the transportability of higher layer protocols.

It is important to understand that the Physical Layer does not contain the medium (e.g. copper conductor, radio waves, etc.) which is used to interconnect systems. This is modelled as being below the Physical Layer and is called the 'Physical medium' as shown in Fig. 4.57. Thus, the Physical Layer is primarily concerned with transporting bits of data from one open system to another using the Physical medium which interconnects the two systems.

The Physical Layer protocol functions will be highly media-dependent and some of this dependence will be visible as variations in the Physical service provided. Thus, there is no requirement to conform to a *single Physical service* as this would unnecessarily restrict open systems to operate across a *single physical medium* and thereby restrict the number of networks which can be used for interconnecting open systems. This would clearly preclude the general model for data communication sought by OSI. Thus,

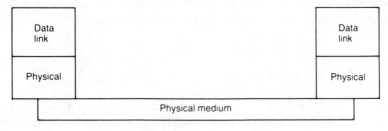

Fig. 4.57 The physical layer model.

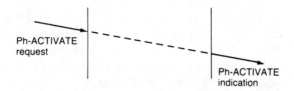

Fig. 4.58 Physical layer activation.

OSI imposes no restriction on the Physical service definition as long as it combines with the Layer 2 and 3 protocols to provide the mandatory Network service.

4.9.2 General requirements

As mentioned in the introduction, the major requirement of the Physical service is to be capable of transporting bits of data from one system to another across a physical medium. This involves the conversion of bit representation from the initiating system format to a transfer format used for transmission and upon receipt a further conversion to the local bit representation of the receiving system. Thus, modems are modelled as Physical Layer functions. None of the many available modems (e.g. V.22, V.24, V.28, etc.) are mandatory *per se* since this would unnecessarily restrict the applicability of OSI. In most cases the modem required is specified by the subnetwork to which the system is attached.

The Physical service must also allow the Data Link Layer to activate and deactivate the Physical connection on which it is the sender of data. It must also be capable of representing full duplex, half duplex and simplex operation, any of which may be imposed by the Physical Layer functions or Physical medium.

Since the Physical service is determined primarily by the functions present in the Physical Layer the following sections will show how various functions which may be present in the Physical Layer can be modelled. These sections also show where a specific service definition for a particular group of Physical Layer machines connected by a single Physical medium may differ.

4.9.2.1 Physical Layer activation

This service allows the user to activate the data transfer mechanism in one direction such that the initiator can send data. This service is unconfirmed so only the Request and Indicate primitives are used (see Fig. 4.58).

This means that there is no indication of the acceptance of the physical connection by the remote Physical Layer and the Physical service user is assumed to recover from the situation arising from all activation failure. By separating the activation of the data channel in each direction the service is capable of modelling the services provided by underlying protocols having duplex, half-duplex and simplex characteristics. No parameters are allowed with the activation primitive.

4.9.2.2 The Physical service seen by one user

The Physical service as seen at one access point is defined in terms of service primitives and valid sequences of primitives which allow certain primitives to be used only according to the previous primitive actions at that service access point. The valid primitives are listed in Table 4.1 and the valid sequences of primitives can be illustrated by use of a state transition diagram for the service boundary (see Fig. 4.59 and 4.60). It is important to note that this is not a protocol description but a method of determining valid primitives in a particular physical service boundary.

Table 4.1 Physical service primitives.

Phase	Service	Primitive	Parameter
Connection Activation	Connection Activation	Ph-Activate Request	None
		Ph-Activate Indication	None
Data Transfer	Data Transfer	Ph-Data Request	Ph-User Data
		Ph-Data Indication	Ph-User Data
Connection Deactivation	Connection Deactivation	Ph-Deactivate Request	None
		Ph-Deactivate Indication	None

4.9.3 Full-duplex, half-duplex and simplex operation

A physical medium, together with the mechanism used to access it, may exhibit one of the following characteristics:

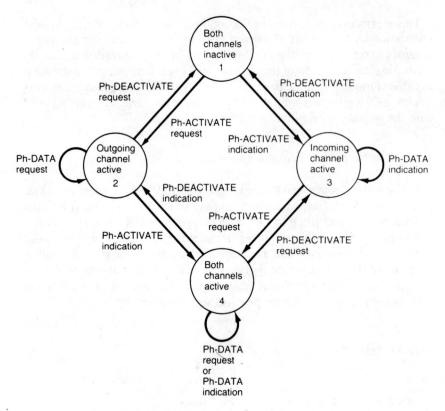

Fig. 4.59 Full duplex state transition diagram.

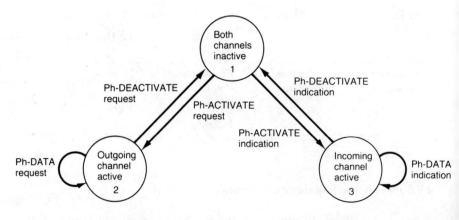

Fig. 4.60 Simplex state transition diagram.

(a) full-duplex, i.e. data can be transferred in both directions simultaneously;

(b) half-duplex, i.e. data can be transferred in both directions but can only be transferred in one direction at a time and the direction of data flow alternates;

(c) simplex, i.e. data can only be transferred in one direction at a time and the data flow does not alternate during an activation.

Only the user of a physical connection can know when it has finished transmitting or requires a response and thus only the user can decide when to pass control of a half duplex connection to the remote user in order to allow it to respond. Thus, a specific Physical service would differ according to these characteristics of the physical layer providing the service. These different characteristics can be described in terms of different valid sequences of the same primitives and different ways of using the primitive sequences.

4.9.3.1 Full duplex

Here the underlying layer is capable of transmitting data in both directions simultaneously so there are no restrictions on the issuance of activation primitives. However, if certain modems are used to provide this service they cannot establish each direction independently, i.e. carrier detection in the reverse direction is required before the initiator is allowed to transmit data. This situation can be modelled as a provider-initiated Activation and result in the issuance of an Activate Indication primitive at the initiator service boundary (see Fig. 4.61). The reverse direction is now established and if the remote user issues an Activate request it is processed locally to the terminal but no externally visible action results.

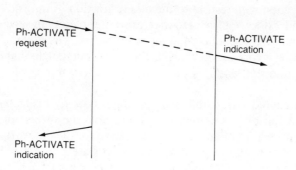

Fig. 4.61 Full duplex activation.

4.9.3.2 Physical Layer deactivation

Like Activation, the Deactivation service concerns itself only with one direction of transmission. Thus, a user-initiated disconnect would have two primitives associated but unlike the higher layers the issuer of a Disconnect request must remain capable of receiving data from the physical connection in the reverse direction if a two-way connection had been established. By the same reasoning, a provider-initiated Disconnect will generate an indication primitive only at the boundary to the Physical service user receiving data from the connection direction which was broken (see Fig. 4.62).

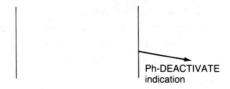

Ph-DEACTIVATE
indication

Fig. 4.62 Physical layer deactivation.

4.9.3.3 Data transfer

Several basic forms of data transfer can be used by the Physical Layer. None of these need be made explicitly apparent at the service boundary. The data transfer service uses two primitives and has an associated parameter which contains the user data to be transferred. The size of the parameter will vary between specific service definitions from 1 bit upwards. In general, the size will be small in order to maximise the efficiency of error recovery mechanisms in higher layers.

Data transfer can only take place following an activation of the connection for that direction of transmission but is independent of the state of the connection with respect to the other direction of transmission.

The provider does not signal to the transmitter of data the occurrence of a provider-initiated disconnect and so the Physical service user may attempt to send data on a connection which cannot transmit data in that direction. Such data will be lost and it is left to the service user to recover from such losses.

In order to avoid such complex modelling situations it is recommended that where a full duplex connection is available the reverse direction is activated upon receipt of an Activate indication for the outward direction. Thus the service would proceed quickly from state 3 to 4 (see Fig. 4.59) hence avoiding the complexities of one user seeing a service in state 3 and the other user seeing a service in state 4.

4.9.3.4 Half duplex

Here the valid primitive sequences are the same as for simplex and hence can be derived from Fig. 4.60. Unlike simplex, however, the use of these primitives will always involve sending information in both directions and, hence, all the valid primitive sequences will be allowed at each Physical service access point.

The major difference between this service and simplex is in the way it is used. There is a commitment on the part of the transmitting user to deactivate the connection at regular intervals to allow the receiving user a chance to transmit data if it has any to transmit. There also exists an agreement that at some time each user will transmit data (this may be only a layer 2 acknowledgement treated as data by layer 1).

4.9.3.5 Simplex

This class of service will allow only one user to transmit data at a time.

In order to maximise the commonality of the service definitions this class of service uses the same set of primitives as the full duplex class of service and changes the service provided by imposing a different sequence on these primitives. The valid sequence of primitives for a simplex class of service can be obtained from the service state diagram given in Fig. 4.60.

Comparing this to Fig. 4.59 the difference, i.e. the lack of state 4, shows the difference in valid primitive sequences. After issuing an Activate request, the user can issue several data requests followed by a Deactivate request. During this time no Indication primitives will be issued by the service provider. Equally, following the receipt of an Activate indication, a user will not issue any request primitives until a Deactivate indication is given.

For simplex operation there is no obligation for a particular user to transmit at all, i.e. the user could always receive on this connection and the service provided may then be optimised to allow only transition involving states 1 and 3. Equally, a transmit-only user may operate over a Physical service which only allows transition involving states 1 and 2.

4.9.4 Functions provided by the Physical Layer

Synchronisation is the process whereby the receiver of data synchronises the times at which it observes the physical medium (to determine whether a 0 or 1 is present) to the times at which the transmitter is manipulating the physical medium (to cause a 0 or 1 to be transmitted).

This is quite a complex subject area in its own right. We will merely observe that there are two common means of achieving synchronisation:

(a) synchronous transmission: whereby one entity generates a continuous regular alternating signal, called a 'clock' signal. Both entities then use this clock signal to determine when to transmit/receive data bits.

(b) Asynchronous transmission whereby a small sequence of bits (called a byte or character) is transmitted as a group. Between byte transmissions, the medium is not manipulated in any way. The receiver is alerted to the arrival of a byte by a 'start' bit which changes the state of the medium. The receiver then samples the medium at regular intervals after the start bit to detect the bits of the byte. This method relies on the availability of separate clocks at each end which are sufficiently accurate that the sender and receiver can stay synchronised for the duration of one byte. This method is in very common use for simple terminals, printers, etc.

The synchronisation function is contained wholly within the Physical Layer and is not visible at the service boundary. In this way, the Physical Layer can be asynchronous or synchronous in operation and the user need not know which.

Certain characteristics of the Physical service provided may, however, be influenced by these Physical Layer functions. If, for example, the underlying protocol was asynchronous and provided for the transfer of 8-bit blocks of data as bytes then this would be reflected in the size of the Physical service data unit. Thus, if there was a significance to the data in 8-bit blocks, e.g. an ASCII character with parity, then these blocks would be transferred in a sensible manner and not split arbitrarily between two blocks. Another example is where the Physical Layer is synchronised and must have some data to send at all times. Thus, a local signal requesting data would be used by the Layer but is not defined at the service boundary. Such a signal would stimulate the user to issue a Data Request primitive which may contain an 'idle signal' (e.g. HDLC flag) to be transmitted.

4.9.4.1 Multiplexing

In addition to this synchronisation and the translation of information for transmission mentioned earlier (e.g. modems) the Physical Layer may provide multiplexing. This allows several logical physical connections to be created using only a Physical medium connection.

A good example of this is ISDN where the Physical Layer protocol is defined in CCITT Recommendation I.430. Here, the use of a multiplexing frame allows three logical connections (one D and two B channels) to be independently available to the Physical service user using only one physical medium connection. Such a frame must not be confused with layer 2 HDLC frames which are totally different in nature. The Layer 1 frame is bit-synchronised, frame-synchronised and must be transmitted repeatedly without a break whilst the Physical Layer is active.

CHAPTER 5
Transport Layer

As we have seen from Chapter 4, the Layers 1 to 3 provide the means to address NSAPs and pass data between them. The nature and quality of the service provided to the Network service user varies according to the characteristics of the underlying medium (or media) in use and according to the use made of it (them) by these lower layers. In particular, the Network service may be connectionless or connection-mode.

Above the NSAP sits a Network service user which is a Transport entity. Transport entities make use of the services provided by the lower layers, described conceptually by the Network Service Definition (ISO 8348) and its Addenda, and build upon them to provide to the Transport service user (i.e. the Session entity) the type and quality of Transport service required.

Specifically, the purpose of the Transport Layer is to relieve the higher layers of all concern as to how to achieve the required service at minimal cost.

When these aims were first set down for the Transport Layer as the Reference Model was being developed, everybody had a good idea what they meant. However, detailed discussions within the group defining the Transport Layer revealed that no two people had quite the same idea as to what exactly was 'cost' or how to minimise it.

Manufacturers believed that cost should mostly refer to cost of implementation together with amount of code needed to be present within a computer.

Users felt that although these aspects were important, cost should also take into account the day-to-day running costs, e.g. as incurred through the use of communications resources, and that the latter were more important.

Failure to agree on how to define 'cost' meant that there could naturally be no technical agreement on how to optimise costs.

However, the current Transport protocol specification does contain sufficient flexibility that broad categories of usage can be 'optimised'. In fact, some of the later proposals for improving the standard are orientated towards making most efficient use of satellites, and further refinements for optimisation in other environments can be expected in the future.

To return to the aims and objectives of the Transport Layer, then, they can be restated to be to provide functionality which satisfies a number of broad categories of application requirements for the transfer of data in a reasonably efficient manner over a variety of (combinations of) subnetworks.

We have mentioned briefly that the Transport Layer is required to provide the user's desired Quality of Service (QOS). To illustrate this further, let us examine the PSTN (Public Switched Telephone Network) and the emerging ISDN (Integrated Services Digital Network). Both offer similar facilities to the user in terms of being able to set-up a switched circuit dynamically, pass data transparently in both directions and clear down the circuit. The QOS available on each is quite different, however. The telephone network can only support data rates of 9.6 Kbps but the ISDN primary channel will operate at 64 kbps. Also, the bit and block error rates are expected to be much lower for ISDN than those currently experienced on the PSTN. Furthermore, the QOS achieved on a network, can, in some circumstances, depend on the facilities requested from the subnetwork when the connection is set up. For example consider a data packet switching subnetwork based on the CCITT Recommendation X.25. Choice of packet size (the maximum size packet which may be transmitted over the connection) and window size (the number of unacknowledged messages outstanding at any one time) can vary the throughput achievable and the end-to-end delays experienced on the connection.

In addition to the quality of Network service issues just described, there are also considerations of the QOS needed by an application. An Application which requires to transfer very large files from one machine to another will probably be more concerned to achieve high bandwidth and low error rates than getting low end-to-end delays. A process control application which receives sensor data and as a result may need to issue 'commands' to alter machinery settings will want low delays but will not be so concerned about throughput. A telex-type messaging service may be more tolerant of errors because of the redundancy in human language which makes deciphering of messages possible even when they contain errors.

Thus, the action taken by the Transport Layer will depend on the degree of mismatch that exists between the QOS needed by the application and the QOS available from the Network Layer.

In order to make the Network and Transport services independent of the underlying media and technology, the service users cannot be required to request explicitly the use of any particular subnetwork or the facilities to be used on the subnetwork. The Transport Layer and Network Layer, however, need to know what is required by the user in order to make those very choices and others affecting operations within the Layer. The model of 'service' that is used is therefore one in which the user specifies requirements for QOS in a network-independent manner by means of parameters such as throughput, error rate, end-to-end delay and others which will be described later.

The standards are also written to assume that an implementation may be very comprehensive, i.e. may simultaneously have to support a number of users, each with different requirements from each other, and may be attached simultaneously to a number of subnetworks, each offering a different QOS. In this complex case, the implementation will have to make dynamic optimising choices internally. The standards contain all the parameters and features necessary for such an implementation to be built. An implementor does not have to implement to such complexity, however. Dedicated systems can be built supporting only a single application and designed to operate over only one type of subnetwork. An example of this would be a Teletex terminal which only offers CCITT Teletex services using the CCITT Teletex protocols over an X.25 network. Such implementations would have such considerations as choice of subnetwork facility and choice of options and parameters within a protocol predetermined and built-in. In this case, the implementation behaves just like a more complex one which has happened to make a single set of choices and which will refuse all attempts to connect to it for other purposes.

Looking now at the history of the Transport Layer standards, work started in ISO on the Transport Layer at the meeting held in Gaithersburg, Maryland, USA in late 1979 and the first two standards were agreed to be ready for publication by the delegates present at the Copenhagen meeting in mid 1984. These two standards have now been published by the ISO Central Secretariat and became available in mid 1986.

These two standards, which were in fact the first two OSI standards to reach full International Standard status, are the Basic Connection-mode Transport Service, ISO 8072, and the Basic Connection-mode Transport Protocol, ISO 8073.

As with any ambitious project with broad scope, there is inevitably follow-on work in the form of enhancements, corrections (although all those known so far are very minor in nature) and addenda.

The rest of this chapter will describe all the major documents and work in progress as at June 1986 relating to the Transport Layer. We will do this in historical order, i.e. by examining the connection-mode documents first and then proceeding to look at connectionless documents and addenda.

5.1   TRANSPORT SERVICE

This section will look at the 'service' provided by the Transport Layer to its user, the Session Layer. The abstract nature of a service definition has already been described previously in this book (Chapters 1 and 3) and the broad objectives of the Transport Layer have been discussed above.

Thus, it will be clear that the Transport service definition objectives are to describe in an implementation-independent way the essential properties and features deemed necessary to permit transparent data transfer and to allow the Transport Service user to express Quality of Service (QOS) requirements for that data transfer.

It will be clear that connection-mode services will be different in many respects to connectionless services.

5.1.1 Connection-mode service

The essential features of the connection-mode service are:

(a) connection establishment, including:
   (i) facility and parameter negotiation;

(b) data transfer with:
   (i) flow control
   (ii) expedited data flow;

(c) connection release.

Flow control is the service that allows the receiver of data to tell the sender when it is unable to keep up with the rate of data transmission, and more will be said of this later.

Expedited data is data which is not held-up by the normal flow control. This means that small amounts of 'urgent' data can be sent and will arrive quickly even when normal data flow is held up because the receiver has become overloaded.

In fact, the intended use of Expedited data is an option which the Transport service users can select and negotiate between them. The use of Expedited data is made an option so that the CCITT Recommendation T.70 (Teletex), which does not provide expedited data transfer, could be incorporated into the Transport protocol. ISO 8072 states that the Transport Layer service provider *must* be able to provide expedited data transfer if it is requested. An implication of this is that the Transport protocol class 0 (identical to T.70, see Chapter 5.2, Transport protocol for an explanation) on its own cannot provide the full Transport service. Only in the special case where it is known that the application will never ask for expedited data transfer will class 0 be sufficient.

In the description that follows, the following terms will be used:

service-user: the entity in the layer above a service boundary which makes use of the services provided by the layer below that service boundary.

service-provider: all the entities below a service boundary which act together to provide the services at that service boundary.

TS: Transport service.

TS user: the Transport service user.

Calling TS user: the TS user which initiates a Transport connection establishment request.

Called TS user: the TS user with which the calling TS user wishes to establish a connection.

Sending TS user: a TS user that acts as a source of data during the data transfer phase of a connection.

Receiving TS user: the TS user that acts as a sink of data during the data transfer phase of a connection.

Some points to note are:

* The terms 'calling' and 'called' apply with respect to a single connection. Overtime, any TS user may act as both calling and called TS user.

* The terms 'sending' and 'receiving' TS user apply to the transfer of a single item of data (Transport Service Data Unit). Over time, both TS users joined by a connection may act as sender and receiver.

5.1.1.1 Connection establishment

Before data can be transferred, a connection must be established between two TS users. The primitives used for connection establishment are:

(a) Connect Request (T-Connect request);

(b) Connect Indication (T-Connect indication);

(c) Connect Response (T-Connect response);

(d) Connect Confirm (T-Connect confirm).

A connection is established by the following sequence:

(a) A TS user, called the 'calling TS user' decides to initiate a connection and as a result issues a T-Connect request primitive down to the Transport Layer.
     At this time, the calling TS user specifies, via parameters in the T-Connect request, the transport service access point (TSAP) address of the remote Transport user entity to which the connection needs to be made (the called TS user), the QOS which is required for the connection and/or whether it is intended that Expedited data transfer services will

be required. This latter is the only optional service element in the Transport service. QOS is optionally specified to the Transport service in terms of the desired, or 'target' requirements and the 'minimum acceptable' levels.

QOS and the use of expedited data will be negotiated as will become apparent.

(b) The Transport service provider receives the calling TS user's request and decides whether it can be accepted or not. If it cannot be accepted, the service-provider rejects the request by issuing up to the calling TS user a T-Disconnect indication.

There can be many reasons why the service-provider cannot accept the request. Some are:

* The minimum acceptable QOS specified in the request cannot be provided because there is no suitable (set of) subnetwork connections available.
* The request cannot be provided at this time, because the provider has too many other connections or because the network is unavailable, etc.
* The requested TSAP address does not exist.
* The called TS user identified by the destination TSAP address does not respond.

(c) If the Transport service provider can accept the request, it is passed to the called TS user who examines it and decides whether it can be accepted or not. If not, the called TS user issues a T-Disconnect request to the Transport service provider which is passed back to the initiator as a T-Disconnect indication.

(d) If the called TS user can accept the request, it will issue a T-Connect response to the Transport service provider. This response may confirm the intended use of Expedited data transfer or may indicate that expedited data will not be supported by the called TS user. It may also indicate the level of QOS that the responder can support, which may be lower than or the same as that proposed in the T-Connect request by the calling TS user but which cannot be lower than the minimum acceptable level.

(e) The Transport Service provider passes the T-Connect response back to the initiator as a T-Connect confirm. In doing so it may further reduce the QOS to a level which can be supported by the provider, again not below the minimum acceptable level.

At the end of this sequence of events, the Transport connection will either have been established with a QOS acceptable to calling TS user, called TS user and the Transport Layer, or will have been rejected by the Transport Layer or the called TS user.

5.1.1.2 Data transfer

Once a Transport connection (TC) has been established, the two TS users may transfer data and expedited data in either direction. For both types of data, the transfer of data is 'unconfirmed'. The sender issues a request for data to be transferred and the receiver gets an indication that the data has arrived, together with the data. No other interactions take place.

The amount of data that the TS user may issue with one service primitive (T-Data request) is defined in ISO 8072 to be unlimited. This means that the Transport service itself imposes no limitation. In any real implementation there will be effective limitations, if only because there is a limit to the amount of data that is actually held on a given computer at any one time. There is a limit on the amount of expedited data that can be sent in one request and this is 16 octets.

The lack of limitation on the size of a T-Data request poses some important considerations for implementations. The service, like all OSI services, is described in terms of primitives which are indivisible and are transferred across the service boundary 'instantaneously'. Clearly, no real implementation can deliver an arbitrarily large amount of data across an interface in zero time. If an implementor decides that his implementation will, in fact, transfer data in discrete, finite amounts (perhaps to fit in with buffering policy) and that therefore a very large data unit will be moved across the service boundary in several physical data units, consistency with the Transport service definition can only be maintained by careful consideration of the fact that the data primitive has not, in fact, been transferred until the last physical item of data has been transferred.

In fact, there will be some local interaction between the sender and the Transport service provider concerned with flow control. The nature of this interaction is not precisely defined in the service definition because it is impossible to be precise without over-constraining implementations. It is merely stated that the Transport service provider may indicate at times to the TS user that further transfer of data is (temporarily) forbidden. More will be said on this subject in section 5.1.1.4 on the Service Model.

As stated before, Expedited data can be transferred if the ability to do so has previously been negotiated during connection establishment. Expedited data, as its name suggests, is expected to be 'urgent' and travel 'quicker' than normal data. By this we do not mean that it travels faster than light! It is only expected that where normal data is experiencing delays within the Transport Layer, Expedited data can potentially bypass those delays.

In fact, the specification of the properties of Expedited data in ISO 8072 sound very weak. The only firm statement that is made is that normal data sent after an item of Expedited data will never be delivered before that

Expedited data. There is, within the model of the Transport service described below, an indication that Expedited data can overtake normal data but it is stated that the amount of overtaking cannot be predicted.

The planned uses of Expedited data known to the author at this time all come under the general heading of interruptions to normal operations. An easily understood example of this is where a user sitting at a terminal decides during some lengthy operation which he has instigated that he no longer wishes to wait for the results but would instead prefer to interrupt that operation and move on to something else.

An implementation which makes no real effort to give expedited treatment to Expedited data and satisfies only the minimum requirements that it should not be overtaken by normal data will work correctly but will not make the terminal user very happy because his request for an interrupt will sit in queues and take its turn with all the other items to be processed. Thus, effective implementation of Expedited data is a matter for efficiency rather than correctness of operation.

5.1.1.3 Connection release

The connection release service provided by the Transport Layer is immediate and can be destructive. That is to say, when release of a connection is requested, any data and expedited data which have not already been delivered may be destroyed. It is a higher layer responsibility to ensure that all activity has been successfully completed before the release of the Transport connection is requested.

There are two forms of connection release – user initiated and provider initiated. The primitives used are:

(a) T-Disconnect Request;

(b) T-Disconnect Indication.

When the TS user initiates connection release, it issues a T-Disconnect request. The remote TS user eventually receives a T-Disconnect indication and the connection is then released (no confirmation is passed back to the TS user that initiated the release).

The Transport service provider itself may initiate release for a variety of reasons such as an inability to maintain the connection with a level of QOS above the minimum acceptable specified when the connection was established.

In that case the Transport service provider issues a T-Disconnect indication to both TS users.

5.1.1.4 Quality of Service

In this section we will look at the qualities of Transport Service which are identified in ISO 8072. We have already noted that QOS is defined in terms of a number of parameters which the TS user can use to specify the requirements from a connection.

The QOS parameters specified in ISO 8072, which are to be refined in a later addendum which is currently subject of study in ISO, are:

* TC establishment delay – allowing the TS user to indicate the maximum tolerable delay between the issuing of a T-Connect request and receiving the corresponding T-Connect confirm. The only real use that can be made of this parameter is to choose to multiplex the connection on to an existing Network Connection (NC) or to choose the appropriate network technology and/or route, these latter being Network service issues.

* TC establishment failure probability – allowing the TS user to specify the reliability needed in setting-up the connection. This can be important for alarm systems, for example. The service provider can only use this parameter in the same ways as establishment delay.

* Throughput – allowing the TS user to specify the throughput required on each direction of the TC. Throughput is only meaningful in the context of attempted transfer of a sequence of service-data units because the effect of lack of sufficient throughput capability is to cause flow control to block issuance of the following data transfer request for a period of time. Clearly, the service provider can use this parameter in a number of ways. It must be taken into account when making Network Layer choices such as route, Network technology and choice of Network parameters, etc. but it must also be used when determining whether a TC can sensibly be multiplexed with others on to an NC. Finally, the service provider can choose to increase throughput on a TC by splitting the transfer of data units over several Network connections.

* Transit delay – allowing the TS user to specify the maximum acceptable delay between issuing a T-Data request and the receiving TS user receiving the corresponding T-Data indication. The service provider can use this to choose the size of data unit used in the Transport protocol and to specify the requirements for Network Layer options as above.

* Residual error rate – allowing the TS user to specify the acceptable rate of *undetected* errors from the Transport service. This parameter is used to determine target requirements from the Network service and also to choose the most appropriate class of Transport protocol.

* Transfer failure probability – allowing the TS user to specify the

acceptable probability that data transfer will not be accomplished to the QOS specified by throughput, transit delay and error rate. The only use this can be put to is in the selection of Network parameters, routes and technologies.

*   TC Release delay – allowing the user to specify the maximum acceptable delay between issuing a T-Disconnect request and receipt by the remote TC user of the corresponding T-Disconnect indication. Again the service provider can use this to set internal timer values and/or control selection of Network technologies, routes and parameters.

*   TC Release failure probability – allowing the TS user to specify that the connection must be able to be released successfully to the stated probability. Clearly, this must relate to performance of the service provider over a number of connections. It can be used to affect choice Transport protocol class.

*   TC Protection – allowing the TS user to specify security requirements. Since the registration of ISO 8072, the work on enhancement of the OSI architecture to take security needs into account has made progress that makes the current specification of TC protection inadequate and this is one of the QOS parameters which requires further refinement.

The parameter is specified currently as taking one of four values allowing the TS user to choose between one of:

(a) no protection;

(b) protection against passive monitoring;

(c) protection against active attacks;

(d) protection against both active and passive threats.

However, the work on security in OSI, as described in Chapter 14, indicates that more choices should be available to the TS user as to types of protection needed.

*   TC Priority – allowing the TS user to specify the relative priority of this TC to others. It is stated that the priority specification will be used by the service provider to determine the order in which:

(a) TCs have their QOS degraded, if necessary;

(b) TCs are released, if necessary, to recover resources.

Again, work is in hand to review and modify this definition. Currently, ISO 8072 states that a single value is used to express priority. However, it has been proposed that priority should be expressed by three values:

(a) priority to gain a connection;

(b) priority to keep a connection;

(c) priority of data on the connection.

    * Resilience of the TC – allowing the TS user to specify the probability of provider-initiated release acceptable in cases where it is very important to keep a connection once established. The service provider may use this to specify Network service requirements or to assist in choice of Transport protocol class.

5.1.2 Connectionless-mode service

Essentially, there is only one service primitive for the connectionless Transport service – the data transfer primitive. As the chapter on architecture has already described, connectionless-mode operation provides for the simple transfer of an item of data, together with the necessary data and QOS parameters such that only a single operation is needed.

The service primitive is the T-Unitdata request. On delivery, the recipient is given a T-Unitdata indication.

Because there is no establishment phase, there can be no negotiation of QOS. Instead, it is assumed that there is a (non-standardised) local means for the TS user to know what QOS can be achieved. In practice, connectionless data transfer is mostly used in situations where the patterns of communication are known and the QOS achievable to each destination is mostly fixed.

Many of the QOS parameters described above for connection-mode service are not relevant to the connectionless mode. These are not only the parameters specifically referring to connections such as establishment delay, they also include the parameters which can only be measured in terms of a sequence of data units. Thus, the only QOS parameters relevant to connectionless data transmission are:

  *Transit delay.
  *Protection.
  *Residual Error probability.

5.2   TRANSPORT PROTOCOL

5.2.1 Introduction

The Transport protocol must provide the Transport service to users at an acceptable quality, bridging the gap between the quality required and the quality available on the connections obtained through the Network service. Thus, the Transport protocol may sometimes have virtually nothing to do and may sometimes have to provide very complex error-detection and multiplexing operations. As a result, the Transport protocol defined by ISO 8073 has a number of options. The major options are the five protocol classes which are explained fully later. These subdivide the wide range of potential Transport protocol functions into five groupings as described below. In addition, some protocol classes have further options. Before outlining the functions of each class, a brief review of the Transport protocol functional elements, known as elements of procedure, and their usage, is given.

5.2.2 Major functional elements of the Transport protocol

This section describes the independent functional elements which can be present in and used by a Transport protocol class. The grouping of such functions in protocol classes is described in section 5.2.3 below.

The Transport protocol operates by exchanging data in units called 'transport-protocol-data units' or TPDUs. There are several types of TPDU, each used for a specific purpose as described below. Each TPDU is given a short name:

CR TPDU – Connect Request

CC TPDU – Connect Confirm

DR TPDU – Disconnect Request

DC TPDU – Disconnect Confirm

DT TPDU – Data

AK TPDU – Acknowledgement

ED TPDU – Expedited data

EA TPDU – Expedited data acknowledgement

ER TPDU – Error

RJ TPDU – Reject

Below is a description of the elements of procedure:

* Assignment. This is the function of deciding which Network connection or connections will carry the protocol exchanges for a Transport connection. Assignment of itself causes no externally visible actions since it only involves the initiator deciding which

connection it will use. The decision is made according to a variety of factors, such as whether there is a suitable Network connection already established, whether there are likely to be more Transport connections established to the same destination and the cost of providing the required Quality of Service (QOS) using the Network connection. If necessary a new Network connection may be established at this time. The Transport entity must decide what QOS parameters to request from the Network Service if a new Network connection is needed. This may not be directly related to the QOS requested on this Transport connection for a number of reasons. The Transport entity may be aware that it will, in general, have to support a number of Transport connections to the end-system containing the called Transport service user specified in the Connect request and may therefore choose to request a Network connection QOS appropriate to that general requirement rather than the specific requirement of a single Transport connection.

When assignment takes place as a result of a previous network connection failure, the transport entity may (re-)assign Transport connections to any suitable Network connection joining the same NSAPs as the original Network connection provided that the Transport entity is the 'owner' of the Network connection and the Transport connection QOS can be satisfied using the already negotiated protocol class over that Network connection. Unless the NCMS additional procedures are being used, the Transport entity is the owner of the Network connection only if it initiated the establishment of the Network connection.

When a Transport connection is being split across many Network connections, a Transport entity may assign the Transport connection to any additional Network connection joining the same NSAPs provided that it is the owner of the network connection and that multiplexing is permitted on the Network connection.

The responder recognises the (re-)assignment when it receives:

(a) a CR TPDU during connection establishment;

(b) an RJ TPDU or a retransmitted CR or DR TPDU during resynchronisation after failure;

(c) any TPDU when splitting is used.

It may happen that all the Transport connections assigned to a Network connection are released, leaving a Network connection with no assigned Transport connections. It is recommended in this case that only the owner of the Network connection should release it and that release should not be initiated immediately after the final TPDU is sent (because Network connection release is destructive and immediate release could destroy the final TPDU).

* Connection establishment. This involves an exchange of Transport Protocol Data Unit(s) (TPDUs). The initiator of the connection sends a CR TPDU and the responder returns a CC TPDU. During the exchange the choice of protocol class and options to be used are negotiated. The general procedure for negotiation of any option at this time is one of the initiator proposing and the responder selecting. In the case of protocol class, the initiator can explicitly list the options which are preferred and the responder will select the accepted class from the list. In the case of TPDU size and QOS, the initiator proposes target values and the responder may accept any 'lower' value. In the case of QOS, 'lower' does not mean numerically lower but means a value representing a 'lower' quality, i.e. lower throughput, higher transit delay, etc.

The connection is identified by a reference assigned by each transport entity. Thus, there are, in fact, two references for each connection, one associated with each entity. The indicator of the Transport connection selects the reference to be used and inserts it in the CR TPDU. The responder inserts his own reference in the CC TPDU along with the initiator's original reference. A reference can be any 16-bit number except zero provided that it is not already used for some other transport connection or frozen (a state that references can be put into after release of the transport connection).

The CR and CC normally contain the calling and called TSAP addresses but either or both may be omitted when the NSAP address to which the TPDU is sent uniquely identifies the TSAP.

When a class is proposed which operates explicit flow control, an initial credit can be sent to the responder in the CR and similarly an initial credit can be sent to the initiator by the responder in the CC.

When class 4 is proposed, the CR contains a checksum and may propose an acknowledgement timer value.

When any class other than 0 is to be used, the CR and CC can also contain up to 32 octets of user data.

* TPDU transfer. This allows Transport entities to transfer TPDUs of defined format in order to provide Transport Layer functions such as connection establishment. This is accomplished by using the (N)-user data field of Network Service primitives on established Network connections.

In Class 1, when the Network Expedited variant has been selected (see description of class 1 below), then EA and ED TPDUs are transferred in Network-expedited primitives. Otherwise, all TPDUs are transferred in ordinary data primitives.

* Segmenting. This function allows for a large amount of user data passed in a single data service primitive to be broken-down into suitable-sized blocks for transmission via the Network service, each block being sent in a separate TPDU. The remote Transport entity can reassemble the data before passing it up to the Transport service user. Choice of TPDU size can influence transit delays and residual error rates as well as costs (e.g. when using X.25 networks).

* Concatenation and separation. These are functions which allow TPDUs to be grouped together to be sent in a single NSDU. This again allows the size of data unit sent over the Network to be optimised. It is necessary to find some way of restricting the maximum size of an NSDU that will be sent in any given situation so that Transport implementations can control the size of NSDU which can be received. This has been achieved in ISO 8073 by imposing the restriction that no more than one non-data TPDU from each Transport connection can be concatenated into one NSDU, together with just one data TPDU. Each transport entity can thus control the largest TPDU it will receive because it negotiates the TPDU size for each transport connection and can control the number of Transport connections that it will accept on any single network connection.

* Connection refusal. This function allows a transport entity to refuse a CR TPDU, either because it cannot be supported by the Transport entity or because the TS user does not wish to accept the connection. The Transport entity sends a DR TPDU in response to a CR in this case.

* Release. This function allows either Transport entity to release an established connection. There are two variants of this function. The implicit variant is where the Transport connection is released by releasing the Network connection (this is only used in class 0). The explicit variant requires an explicit TPDU exchange to release the Transport connection. The latter allows the Transport entity to recover from loss of the network connection and allows release of just one of the Transport connections in the case where more than one is multiplexed on to a Network connection. It also allows sequential re-use of a Network connection which can permit cost optimisations where connection establishment costs are high and/or reduce TC establishment delays.

* Implicit termination. This is where loss of or signalled errors on the network connection causes termination of the Transport connection without further protocol exchanges.

* Data TPDU numbering. This is a function used in some protocol classes to assist in error recovery and flow control. There are two variants, normal and extended numbering. Extended numbering is

used for networks where the bandwidth is very high and/or the end-to-end delays are long (e.g. a satellite link) and where as a result there may be a considerable number of TPDUs sent and not acknowledged at one time in order to maximise throughput.

* Expedited data transfer. This is the protocol mechanism which supports the Expedited Data primitive of the Transport service. Mostly, the ED TPDU is transferred using the normal data transfer of the Network service. Its ability to be transferred 'expeditiously' therefore depends on the implementation of the transport entity at each end of the connection and the extent to which expedited data will be allowed to bypass queues of normal data and normal data flow control.

* Reassignment after failure. This is the function invoked when trying to recover a transport connection after failure of the network connection to which it was assigned.

* Retention until acknowledgement of TPDUs. This is the function used to support error recovery after possible loss of TPDUs by the network. The sender of data retains a copy of that data until an acknowledgement is received from the remote transport entity. The acknowledgement signifies that the sender will never be asked to retransmit that TPDU and so it may be safely discarded.

* Resynchronisation. This is the function invoked after network signalled Reset or Disconnect to restore the transport connection to normal. It is only used in classes 1 and 3. Class 4, which can also recover from network failures, uses different mechanisms based on time-out acknowledgements of each TPDU sent.

* Multiplexing and demultiplexing are the functions used to enable more than one Transport connection to be carried over a single network connection. This function can be used by the Transport Layer in some circumstances to optimise costs by minimising the number of network connections required to support an application.

  Note that cost savings cannot always be achieved by multiplexing. When using an X.25 network, for example, all 'user data' transferred over the network connection is charged for. The multiplexing procedures within the Transport protocol use explicit flow control (see below) which required sending acknowledgements for received data back to the sender as user data over the network connection. This can therefore increase costs by an amount which depends on the extent to which acknowledgements can be concatenated with data sent in the return direction (not often achieved in most implementations).

* Explicit flow control is a function used when multiplexing and therefore the backpressure flow control exerted by the network

cannot be used to control the flows of each Transport connection being carried. It requires TPDU acknowledgement exchanges and can incur operating costs on some networks such as packet-switched networks (as explained above).

The flow control mechanism uses the concept of 'credit'. For each direction of flow, the receiver will issue to the sender credit which is specified in terms of the number of TPDUs which the sender may transmit and have unacknowledged at that time.

Credit is issued at connection establishment time and may be 'topped-up' during the data transfer phase of the connection when an AK is sent. Unlike the X.25 protocol, however, the acknowledgement of receipt of data is decoupled from the issue of credit. In X.25 the credit, or 'window' is defined to be fully open when the connection is established and is automatically opened up when data is acknowledged.

In ISO 8073 however, the receiver can control the amount of credit issued at start. As with X.25 the credit is reduced by one automatically whenever data is sent but, unlike X.25, credit is a separate field in the AK so that the data can be acknowledged without issuing further credit if desired.

In classes 1, 3 and 4 only, there is a further possibility of credit reduction by the receiver. This is to allow for implementations handling a large number of connections simultaneously where credit is issued to all connections but there are, in fact, not enough resources to cope should all the connections send simultaneously. In these circumstances, the receiver can issue a credit reduction on chosen connections and may choose to discard data already sent if it fell outside the now reduced credit limit by the time it arrived. The error recovery procedures are used to cause retransmission of this discarded data, which is why the facility is restricted to the error recovery classes only.

* Use of checksums is made when the Transport Layer is required to perform error detection because the undetected error rate of the network connection is unacceptably high compared to that required by the TS user.

The sender calculates a checksum and inserts it as a parameter in the TPDU. The checksum defined has been chosen so that when the receiver applies the checksum calculation on the whole TPDU then it should sum to zero. This means that the receiver does not need to examine the structure of the TPDU in order before determining that it has been received without error.

The checksum algorithm has been chosen to be easy to calculate rather than give high protection. It should give adequate protection for normal usage in a situation where the type of error is likely to be the result of malfunction in a network node rather than bit errors on the transmission medium (these latter are assumed to be eliminated by the Data Link Layer).

Thus the use of Transport protocol class 4 with checksums should not be seen as an alternative to the use of Link Layer error detection and recovery over a noisy medium.

* **Frozen references** is the function invoked after release of the transport connection which forbids re-use of the unique reference number identifying that connection for a period of time long enough that there will no longer be any TPDUs in existence belonging to the connection either in the network or in the end-system when the reference is eventually re-used. This prevents possibility of confusion between TPDUs from a new connection and any from previous connections.

* **Retransmission on timeout** is a function used to assist with error recovery over very poor-quality networks which do not signal data loss with a sufficient degree of reliability. In these cases a timer is set running when data is transmitted and if no acknowledgement has been received before expiry the data must be retransmitted.

* **Resequencing** is another function used over very poor-quality networks which cannot be trusted to preserve the sequence of transmitted data, e.g. datagram networks or LANs using multiple MAC bridges. The receiver of data must examine a sequence number and if necessary re-order received TPDUs before passing their data to the Transport user.

* **Inactivity control** is a function used over very poor-quality networks to ensure early detection of loss of communication. In the absence of user data over a time interval a TPDU is sent anyway so that the receiver can assume loss of the connection after a fixed interval if no TPDUs are received. This function can improve resilience but can also incur operating costs.

* **Splitting and recombining** is the function used to spread a Transport connection over a number of underlying network connections in order to provide greater throughput or greater reliability than can be obtained with any single available network connection. It should be noted that whatever the quality of the network connections used, when splitting is used the result is like a poor-quality connection in the sense that data can be mis-sequenced at the receiving end.

5.2.3 Protocol classes

The preceding section reviewed the functions that can be present in the Transport protocol. These functions are not present in every implementation of the Transport protocol and even where they are present they are not always invoked for every user. There are classes of protocol which provide a sequence of increasing level of Transport protocol functionality as described below. In addition, there are options within some classes to provide for minor variations:

* Class 0 is the simple class. It is fully compatible with CCITT S.70 and thus TPDU lengths supported are restricted to 128, 256, 512, 1024 and 2048, i.e. it will not support 4096 or 8192. As with all the classes the default TPDU size is 128 and during negotiation any proposal to use another size can always be negotiated back to 128. It does not provide for transfer of user data during connection establishment or for support of the Expedited data service.

  Class 0 should be used where the qualities of the supporting network connections are adequate to directly support the application and where multiplexing will not be required.

* Class 1 is the basic error recovery class. It provides recovery from network-signalled errors or disconnects and will optionally support Expedited data transfer if the underlying network provides it. The major option in this class is the optional use of network receipt confirmation to acknowledge and release retained data TPDUs. This technique can achieve considerable savings in the number of NSDUs transmitted and as a result reduce operational costs in cases where the network charges are related to data transmissions, e.g. public packet networks.

  Class 1 should be used where the supporting network connection can be trusted to signal errors but where the expected frequency of errors or disconnects is too high for the application and no multiplexing is required.

* Class 2 is the basic multiplexing class. It provides for multiplexing without error recovery. The major option within this class is the optional non-use of explicit flow control on each Transport connection. When it is not used, it must be assumed that each Transport connection is lightly loaded or that some other mechanism will ensure that there is no overload or contention between them. Class 2 will also optionally support extended numbering which permits the use of larger windows for flow control and error recovery purposes.

  Class 2 should be chosen when multiplexing is required for cost or other reasons and the qualities of the underlying network connection are suitable for the Application using this Transport connection.

* Class 3 is the error recovery and multiplexing class. Like class 1 it will recover from Network-signalled errors or disconnections but does not provide additional error-detection mechanisms. The only option in this class is the use of extended TPDU numbering.

| Function | | Class | | | | |
Protocol mechanism	Variant	0	1	2	3	4
Assignment to Network connection		★	★	★	★	★
TPDU transfer		★	★	★	★	★
Segmenting and reassembling		★	★	★	★	★
Concatenation and separation			★	★	★	★
Connection establishment		★	★	★	★	★
Connection refusal		★	★	★	★	★
Normal release	implicit	★				
	explicit		★	★	★	★
Error release		★	★			
Association of TPDUs with Transport connection		★	★	★	★	★
DT TPDU numbering	normal		★	★	★	★
	extended			★	★	★
Expedited data transfer	network			★		
	normal		★	★	★	★
	network					
	expedited	★				
Reassignment after failure			★		★	★
Retention until acknowledgement of TPDUs	Conf. Receipt		★		★	★
	AK		★		★	★
Resynchronisation			★		★	★
Multiplexing and demultiplexing				★		
				★	★	★
Explicit flow control with					★	★
without		★	★	★		
Checksum (use of)						★
(non-use of)		★	★	★	★	★
Frozen references			★		★	★
Retransmission on timeout						★
Resequencing						★
Inactivity control						★
Treatment of protocol errors		★	★	★	★	★
Splitting and recombining						★

Fig. 5.1 Table of transport protocol functions and classes.

Class 3 should be chosen when multiplexing is required but the expected frequency of network-signalled errors is too high for the application.
* Class 4 is the error recovery and detection class. It provides all the class 3 functions plus the ability to detect unsignalled errors and also to detect and recover from mis-sequenced data. It also provides for splitting.

Class 4 should be used when the network undetected error rate is too high for the Application or where splitting is required to increase bandwidth or reliability.

When multiplexing is used on a network connection, Transport connections of different class can be supported on the network connection. Also, when class 4 is being used to split a Transport connection over a number of network connections, it can make use of any available network connection to the desired NSAP (or more precisely to the desired Transport entity in the case where it can be reached through more than one NSAP known to the sender) which is being used for multiplexing, i.e. not supporting class 0 or 1.

See Fig. 5.1 for a summary.

5.2.4 Conformance requirements

The existence of five protocol classes together with options within them introduces a high probability that implementations may not interwork due to incompatible choices of class and options and this would defeat the objectives of OSI. Thus the Transport protocol specification, ISO 8073, contains a 'conformance' statement for implementations. Although all the ISO member bodies agreed on the desirability of having a single class and option set which all implementations should be required to support in order to ensure that any two could interwork they could not agree on a single choice for this. The conformance section thus requires that any conforming implementation must support either class 0 or class 2 with explicit flow control. It also requires that if class 1 is implemented then class 0 must be implemented and if classes 3 or 4 are implemented then class 2 must be implemented.

The result of these requirements is that there are now only two groups of mutually incompatible implementations: those supporting class 0 and those supporting class 2. This is better than a totally uncontrolled situation but still not ideal.

The CCITT version of the Transport protocol, X.224, is significantly different in this one respect. X.224 static conformance requirements state that a conforming implementation *must* implement class 0 *and* class 2.

All implementations are required to be able to support a TPDU size of 128 octets, other sizes being optional.

Apart from the explicit options within the Standard there are also hidden options. Some implementations may be of the type where they can only initiate outgoing connections and others may be of the type where they can only accept incoming connections. Thus, neither sending of connection request TPDUs nor accepting connection request TPDUs can be made mandatory. In order to ensure that purchasers have adequate information to determine that two implementations can interwork (and to alert them to the fact that these details need to be checked) the conformance statement also requires that suppliers provide information stating which classes and options have been implemented.

5.2.5 Protocol providing the connectionless service

A separate protocol standard, ISO 8602, specifies the protocol to be used to provide the connectionless Transport service over a connection-mode or connectionless-mode network service.

The protocol is very simple. The only TPDUs used are the Unitdata (UD) and the UN (see the description of the network connection management subprotocol).

When a T-Unitdata Request is received by the Transport service-provider from the Transport service user, a corresponding UD TPDU is transmitted, either over the connectionless network service in the user data part of an (N)-Unitdata Request or in the user data part of an (N)-Data Request on a suitable connection.

When using a network connection there are two additional concerns.

First, a suitable network connection must be used. A network connection cannot be used to transfer UD-TPDUs *and* to carry Transport connections. If a network connection must be established to convey a UD-TPDU then the (N)-user-data field of the (N)-Connect Request must contain a UN TPDU specifying the use of the connectionless Transport protocol on the network connection.

Second, because the release of a network connection is destructive and can destroy NSDUs in transit and not delivered, the sender of a UN TPDU must set a local timer and not release the network connection until the timer has expired.

The UN TPDU contains the following parameters:

(a) source TSAP ID;

(b) destination TSAP ID;

(c) checksum (optional).

The checksum is inserted by the sender of the UN TPDU if the QOS specified in the T-Unitdata Request cannot be met by the residual error rate provided by the Network service. The receiver must check the checksum if it is present.

5.3   ADDENDA AND ENHANCEMENTS

There are two addenda to ISO 8073 not already described above.

5.3.1 Network connection management sub-protocol (NCMS)

This is a protocol mechanism to give greater flexibility in the use of network connections and re-establishment of failed network connections than is currently possible and also a mechanism for identifying the Transport protocol being used over a network connection.

ISO 8073 specifies that only the owner of a network connection can assign Transport connections to it. This is to ensure that there is no clash over the use of the network connection, e.g. one Transport entity wants to use a class 0 or class 1 connection, neither of which can be multiplexed over the network connection, and the other Transport entity tries to initiate one of the multiplexing classes (2, 3 or 4).

NCMS allows the initiator of a network connection to choose whether to grant to the responder the right to assign Transport connections to the network connection. NCMS provides the NCM TPDU which enables the initiator to signal this choice to the responder.

The NCMS procedures form an optional extension to the procedures of ISO 8073 and were designed to permit interworking between implementations which operate NCMS and those which do not.

The UN and NCM TPDUs are conveyed in the (N)-user-data parameter of the (N)-Connect request.

5.3.1.1 Protocol identification

NCMS provides the UN TPDU to carry a protocol identification parameter. Optionality of use and compatibility with ISO 8073 determines that if the UN TPDU is not present in the (N)-Connect request then the protocol to be used is that defined by ISO 8073 without the NCMS addendum procedures.

The currently specified encoding for this TPDU in the NCMS addendum defines only the code representing the use of ISO 8073.

The protocol for providing the connectionless Transport service defines another encoding for the UN TPDU which specifies the use of that protocol.

The encoding of the UN TPDU is actually a rather devious compromise in order not to clash with the CCITT parameter used to identify CCITT non-OSI protocols such as X.29 and yet to preserve the standard ISO parameter encoding technique. It relies on the fact that CCITT use only 1 octet for their parameter which can take the value 1, 2 or 3. The first octet of the ISO UN TPDU is a length-of-parameter field which will not clash with the CCITT field because the ISO parameter has been made at least 5 octets long!

5.3.1.2. Assignment rights

The NCM TPDU is provided to permit the initiator of a network connection to grant assignment rights to the responder. Because the NCMS procedures are optional, the use of NCMS is confirmed by the responder when it is capable of operating the procedures. If the initiator receives no confirmation of this from the responder then the NCMS procedures will not be operated on the network connection.

The initiator can indicate in the NCM TPDU which entity(s) have the right to assign transport connections as follows:

(a) sender only

(b) RA – responder only

(c) AA – all

NCMS, then, defines that a Transport entity is designated the 'owner' of the network connection if NCMS has been used and the entity has been given assignment rights (if NCMS is not used, then clearly the 'owner' is as described by ISO 8073, i.e. the initiator of the network connection). This widened definition of 'owner' can then be used to permit all of the procedures of ISO 8073 to be followed as written.

The NCM TPDU also requires that the initiator of the network connection should assign an identifier, the NC-reference, to the network connection.

This is used in situations where two end-systems wish to maintain a specific number (more than 1) of network connections between them but wish either end to be able to initiate network connections, i.e. neither is singled-out to be always the initiator. It is also used to identify that a network connection being established is a new one or is a recovered connection after a previous connection has failed. In both these cases, since both end-systems will be trying to achieve the same objective, that of (re)establishing 'N' connections, it can happen that there may actually be up to 2 × N-connection establishment requests in progress at one time. In NCMS parlance, this situation is known as a 'collision'.

In order to handle collisions, it is necessary for each end-system to choose independently to persevere with the same set of 'N' connections and drop the remainder. If each were to pick connections to drop randomly then there will be in general less than 'N' connections remaining and the situation would be unstable.

Collision resolution is therefore based on the quality of network connection established. This assumes that the initiator of each connection can know in advance the quality of connection that will be obtained, which can only be determined in the special cases where there are no network relays that can select alternative routes with greatly different QOS. Where QOS cannot be predicted by initiator (and thus known at both ends of the connection) then collision resolution is based on comparison of the NSAP addresses which is done in such a way that both ends form the same view as to which NSAP address has the lower value. Network connections initiated by the Transport entity operating through the 'lower' value NSAP are retained and the others are released.

The NCMS protocol definition appears to be complex because there are so many cases, e.g. responder is/is not capable of applying the procedures, use of QOS/NSAP addresses to determine which connection to drop when there are collisions, whether the connection is new/recovered. Only the basic principles have been outlined here.

5.3.2 Use of class 4 over the connectionless network service

The class 4 procedures defined in ISO 8073, as with all procedures in ISO 8073, specify only the use of the connection-mode network service for the conveyance of TPDUs. In fact, class 4 contains all the functions needed to be used over the connectionless network service and an addendum is under preparation specifying how to use class 4 in this way.

The changes needed are very trivial, just adding the Unitdata service primitive to the list of network service primitives used for the conveyance of TPDUs.

PART 3

COMMUNICATION BETWEEN OPEN SYSTEMS: COMMON SERVICES

CHAPTER 6

The Session Layer

6.1 INTRODUCTION

At the Session Layer we cease to be concerned with the means to provide data transfer. The Session Layer and higher layers are instead concerned with the provision of common facilities to applications which wish to communicate.

The service provided by the Transport Layer is two-way simultaneous and with Expedited data transfer in both directions. The diagram below shows a possible sequence of data flows on a Transport connection:

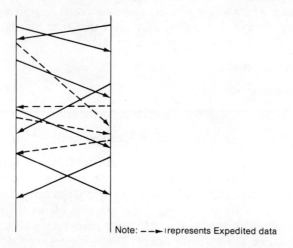

Note: ‒ ‒► ।represents Expedited data

Fig. 6.1 Data flows on a transport connection.

These facilities from the Transport Layer need to be controlled by the application in order for two end-systems to coordinate and make sense out of the data exchanges.

Broadly, the purpose of the Session Layer is to provide structured and synchronised control of the communications channels provided by the Transport Layer.

What this means in practice can be described informally in terms of equivalent functions encountered within human communication.

When two people talk, there are indications given as to when the speaker is willing to allow the listener to speak. These are given by the inflection of the voice and, if the two are face-to-face, by raising the eyes to look at the other person. These clues are often not consciously recognised but it is very clear when the system breaks down. For example, I have often had occasion to make transatlantic telephone calls which have been routed via satellite and as a result there is a half-second delay between when I speak and when the listener starts to hear me. This interrupts the normal conversation mechanisms such that it is very frequent that both parties start to talk at the same time, i.e. dialogue is not being controlled.

Another good example of this is when radio communications are used whereby the radios at each end have to be manually switched from talk to listen and vice versa. In this case the talker has explicitly to hand over the channel to the listener, usually by the convention of saying 'Roger'. The Session Layer manages this kind of situation by use of 'tokens' which must be possessed before data can be sent (saying 'Roger' is equivalent to handing over a 'permission to speak' token).

More formally, a token is defined to be an attribute of a session connection which is dynamically assigned to one session service user at a time. Assignment to a service user gives that user the right to invoke a service primitive. Tokens exist to control:

(a) data transmission;

(b) connection release;

(c) minor synchronisation (see below);

(d) major synchronisation/activity management.

Another phenomenon in human communication is when the listener becomes distracted and misses some of the conversation. When his attention is returned, he sometimes has to say 'I'm sorry, I missed that bit, can you repeat from ...' This is an example of resynchronisation, the purpose of which is to ensure that both parties are at the same point of understanding in the conversation. The Session Layer also provides for resynchronisation after a system has suffered a transient failure.

Finally, it can happen that the subject of conversation can change radically. It is important for the talker to ensure that the listener is made aware when this happens so that the two do not talk at cross-purposes subsequently. The Session Layer provides services such that the application may 'mark' points in the dialogue and the Session Layer then ensures complete separation of the data stream before and after the mark. This can be done either by using 'Major Synch points' or by using 'Activities'.

The following sections will describe the Session Layer services in a more precise way.

Before that, however, it is useful to make the observation that, whereas the Transport Layer has a simple set of services with many optional ways of providing them, the Session Layer has many optional services but with only one way of achieving each one. The options available within the Session Layer are thus all optional *services*, not optional protocol elements and are selected by the application standards, *not* by implementation choices.

A major concern during the development of the Session Layer standards has been the issue of compatibility with CCITT standards, most notably the T-series Teletex recommendations. This has led to two styles of separation of Session-user conversations on a Transport connection and corresponding resynchronisation procedures, namely 'Activities' (CCITT) and 'Major Synch points' (ECMA). Since these services are user-invoked, achievement of Teletex compatibility when using ISO Session services becomes the responsibility of the Session service user. Guidelines have been produced on how to use the Session Layer in a Teletex-compatible manner.

Another major concern during the development of the Session Layer standards was over the style of dialogue control which best-suited application protocol development.

There were two views expressed. One stated that two-way simultaneous data flow was needed for efficient operation, i.e. either end can send data at any time. The other view was that this made application protocols too difficult to design to operate correctly since it gave too many possible collision cases. The alternative approach put forward was that two-way alternate working was needed where each end took turns to transmit data.

As a result the Session Layer supports both styles as alternative options. As indicated above, control of 'turn to speak' in the case of two-way alternate mode of working is by the use of a token.

6.2 THE SESSION SERVICE

The Session Layer service is defined in ISO 8326. Work started in ISO on the Session Layer in 1979 and the IS text was approved in late 1984.

We have seen above that there are a great many optional services in the Session service provided to its users. The resulting proliferation of service primitives are grouped together into 'functional units', each functional unit representing the collection of primitives needed to provide one of the major optional styles of working. Functional units are the subject of negotiation at connection establishment time. In addition, the Session service defines three subsets, each consisting of a number of functional units. Subsets are not negotiated and their only use is to provide convenient human labels for the grouping of functional units. No real use has been made of these subsets and they are therefore of no significance.

The Session Layer services can be subdivided into the usual three main phases of a connection:

(a) establishment;

(b) data transfer;

(c) release.

In addition, the data transfer phase service primitives can be grouped into:

(d) data transfer and dialogue control;

(e) synchronisation;

(f) activity management;

(g) miscellaneous.

6.2.1 Connection establishment

Session connection establishment proceeds in much the same way as establishment in the other layers:

* An initiator (the 'calling Session service user') sends an S-Connect request which is eventually passed to the desired responder (called the 'called Session service user') as an S-Connect indication. The request contains a number of parameters which are described below and for the most part subject to negotiation in the normal way.

(a) The Session connection identifier is a parameter whose value is chosen by the calling service user. It need not uniquely identify the Session connection. Mostly, the service will choose to make use of it during recovery but the usage of the identifier (as with many Session Service parameters) is entirely at the discretion of the service user (i.e. the Application standard). The identifier is passed through to the receiving Application entity by the Presentation Layer.

(b) The called address specifies the Session service user to which a connection is required. The Session service does not specify the content of the address fields (and neither does the Session protocol). These are specified in the addendum to the Reference Model covering naming and addressing (ISO 7498 DAD2). Note that at present the corresponding address in the response primitive must be equal to the called address, thus not allowing for generic addressing or for call redirection.

(c) The QOS parameters specify the QOS required by the Session service user. The QOS parameters are broadly the same as those present in the Transport and Network Layers. There are currently (at the time of writing) small differences in the specification of each parameter and the way that QOS is requested in each layer, however. These are the subject of a multi-layer review in ISO and it is hoped that they will be brought into line with each other.

There are two QOS parameters not present in the lower layers. One (the extended control parameter) relates to the behaviour of the Session service when normal data transfer is blocked by flow-control. This parameter allows the service user to specify that in these circumstances it is not acceptable for other service primitives to be blocked. In fact, this parameter was inserted as a means to indicate that the Expedited data service of the Transport Layer should be provided in an effective manner since the Session protocol uses this in order to send some service primitives. Chapter 5 discusses the effective provision of Transport Expedited data.

The other, the optimised dialogue transfer parameter, permits the Session service provider to concatenate certain SSDUs. How the Session service provider achieves concatenation is a local matter at the sending end. The effect is to invoke a protocol mechanism called the extended concatenation option to transmit concatenated SSDUs.

(d) The Session requirements parameter allows the two Session service users to check that the Session Layer implementations at each end can both provide the set of functional units needed by the application. We have already seen that there are a great many optional services in the Session Layer and as the set of services needed will vary from application to application, so many Session Layer implementations will only implement the subset of services needed to support the set of applications that have been implemented. By indicating the set of services, grouped into functional units, that are needed for the connection at time of establishment, the situation is avoided whereby a connection is established and only later is it discovered that it cannot be used.

ISO 8326 states that the requirements listed in the S-Connect response need not be the same as those in the preceding S-Connect indication, i.e. the responder can specify requirements not received in the incoming connection. However, it also states that only the set of functional units common between the indication and the ensuing response are the ones agreed for use on the connection.

(e) Parameters also exist to specify initial values for synchronisation point serial numbers, and the initial location of tokens. Tokens are either:

(i) unavailable, in which case either end can use the corresponding service at any time;

(ii) requestor side, i.e. located with the initiator of the connection;

(iii) acceptor side, i.e. located with the responder or acceptor of the connection;

(iv) acceptor choose, which allows the acceptor of the connection to choose which side should hold the token.

As stated above, there are tokens for data transfer, connection release, synchronise (minor) and synchronise (major)/activity. The location of each of these tokens is specified independently.

* The Session service provider may reject the connection, in which case it passes an S-Connect confirm back to the initiator with the result parameter indicating provider-reject and giving a reason for the rejection.
* If the provider does not reject the connection, the responder accepts or rejects the connection. If the connection is rejected an S-Connect response is issued by the responder with the result parameter indicating rejection with a reason for the rejection and this is passed to the initiator as an S-Connect indication.
* If the responder accepts the connection it issues an S-Connect response primitive with the result parameter indicating acceptance.

6.2.2 Dialogue control and data transfer

As discussed above, there are two styles of dialogue and data transfer control, the duplex (two-way simultaneous) and half-duplex (two-way alternate) styles. In the half-duplex style, the transfer of data is conditional on possession of the data token.

We have also seen that there can be a need to partition parts of the dialogue and there are three mechanisms for doing this:

(a) minor synch points (suitable only for one-way data flow with no use of Expedited data);

(b) major synch points (suitable for general dialogues);

(c) activities (suitable for two-way alternate with no transfers between activities).

For each of these services, the Session service defines the restrictions on their use, their interactions and the precise state that the connection will be set to when issued and, more importantly, when resynchronisation occurs. The relevance of this to applications is entirely dependent on how the Application standard chooses to use the services. For example, the Teletex application, as defined by CCITT, uses the activity service to separate documents on a connection.

On the other hand, it is proposed that future Common Application Service Elements (CASE) might use major synch points to mark a change in Application context, e.g. from file transfer to message-passing.

The use of each of these services is negotiated at connection establishment time. When each is available to be used, the appropriate token is also designated to be in use and its initial location defined as described above.

6.2.2.1 Token management

There are three token management primitives:

(a) S-Token-Give: this primitive is used to pass one or more listed tokens to the receiver. Naturally it can only be used to give tokens which are in the possession of the sender.

(b) S-Token-Please: this primitive is used to request that the receiver give one or more listed tokens to the sender. Up to 256 octets of user data can also be sent with this primitive. Naturally, only tokens which are not possessed by the sender can be requested.

(c) S-Control-Give: this primitive is used by the sender to give all tokens in its possession to the receiver. In fact, this primitive is equivalent to one of the CCITT Teletex primitives and because of its origin is regarded as part of activity management. It is only available if the activity management functional unit has been requested.

6.2.2.2 Data transfer

Normally, user data is transferred using the S-Data primitive. The sender issues an S-Data request with user data (unlimited in maximum length but greater than zero octets) as its only parameter. This is passed to the receiver as an S-Data indication.

The Session service also provides for Expedited data transfer. As explained in Chapter 5, Expedited data does not necessarily travel faster than normal data but should be able to be conveyed even when normal data is subject to flow control and cannot (temporarily) be conveyed. The S-Expedited data primitive allows from 1 to 14 octets of session user data to be transferred.

There are two further types of data transfer primitive, S-Typed-Data and S-Capability-Data and these are discussed below.

It is recognised that Session user data comprises two distinct types:

Application user data and protocol data units from Layers 6 and 7. It is further recognised that for some applications the latter should not be restricted to the token control which is exerted over the former. Token control exists to manage the dialogue between two applications, but correct functioning of Layers 6 and 7, especially during error or recovery situations, may well require unrestricted protocol exchanges.

Thus, the S-Typed-Data primitive has been provided. Data can be passed using this primitive regardless of the location of the data token. As with normal data, this primitive can carry unlimited amounts of user data. Although originally conceived as needed only for the half-duplex functional unit, the service was eventually agreed to be provided for the full duplex case as well. This can allow greater consistency in the design of higher layer protocols since they may choose to pass their protocol data units using S-Typed-Data at all times, regardless of whether the application dialogue is being constrained to half duplex or not.

Note that S-Typed-Data is subject to the same flow control as normal data and so if one is blocked then the other is blocked also. Thus, when S-Data and S-Typed-Data are blocked, only S-Expedited-Data can be passed.

Finally, the S-Capability-Data primitive has been provided to allow a limited amount of user data to be transferred when the activity management functional unit has been selected but no activity is in progress. The S-Capability-Data primitive can carry an unlimited amount (previously limited to between 1 and 512 octets) of user data. It is subject to token control, however. The Sender must possess the major/activity token and, if they are in use, the data and synch minor tokens as well.

6.2.2.3 Synchronisation services

As explained in the introduction to this chapter, the Session Layer provides various means to mark points in the dialogue and to resynchronise the two Session users after failures. Note that the lower layers provide the means to recover transparently from short-term transient failures in the communications pipe. The Session Layer services are more orientated to aiding recovery after a long-term failure of the communications pipe or a temporary failure of one of the end-systems themselves. In the latter case, it must be assumed that some state information and data may be lost by the end-system.

Resynchronisation services of the Session Layer are based on the 'marks' placed in the dialogue. These marks are set by the Session service users at intervals decided by the Application. There are two types of mark, the minor synch points and the major synch points. The minor synch points define marks to which recovery back to a previous state can be requested. Major synch points also mark the dialogue but in addition divide the dialogue such that recovery using the 'restart' resynchronisation option (see below) cannot take place to any mark that occurred before the major synch point.

Minor synch points need not be acknowledged by the receiver and data, typed data, Expedited data and Capability data can all be sent by the sender and receiver of the minor synch point before an acknowledgement is sent or received. As Expedited data can 'overtake' other primitives, it is possible that the two Session service users will not have the same view of what was sent before the minor synch point and what was sent after it. Depending on the way that the Application standard operates and uses the Session services, this may or may not be significant.

Where the Application standard does make use of the various different forms of data transfer provided and does wish to have a clean separation between data sent before and after a mark, the major synch point must be used. Practically no other primitives can be sent after a major synch point until the corresponding acknowledgement has been received. Thus, resynchronisation based on major synch points will be more deterministic in the general case. This is obtained at the penalty of interruption of normal data flow whilst waiting for the acknowledgement.

Before the detailed description of synch points and their use it will be useful to consider session dialogue separation and some of the problems.

6.2.2.4 Session dialogue separation

Figure 6.2 shows two sorts of 'dialogue' (messages between the two session users). Dialogue 6.2(a) is very simple – a one-way flow of data using only the Transport normal flow.

A very simple session PDU can be inserted in this flow (see Fig. 6.2(b)) with the property that it identifies a point A at one end and a point B at the other end such that all events prior to A at one end correspond to events prior to B at the other, and similarly for after A and B. There is a simple and clear separation of dialogue P1 (prior to A and B) from dialogue P2 (following A and B).

Now consider the fully general dialogue Fig. 6.2(c), with two-way transmission and expedited messages overtaking normal flow. If we want to separate the dialogue into two, how can we do it? We start at C on one side, but where is the corresponding point on the other? If we look at each of the four flows (each way and normal/expedited) we get four different points at the other side. Clearly, a more complex protocol exchange will be required. This is the problem *major synchronisation* has to solve.

The easy one:

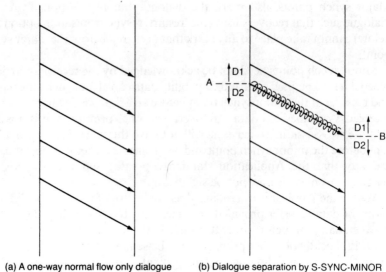

(a) A one-way normal flow only dialogue   (b) Dialogue separation by S-SYNC-MINOR

The hard one:

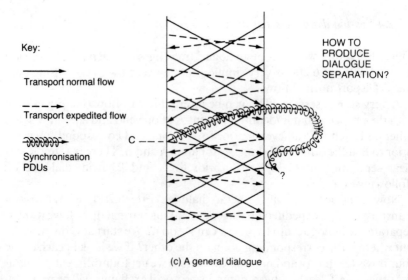

(c) A general dialogue

Fig. 6.2 Session dialogue separation.

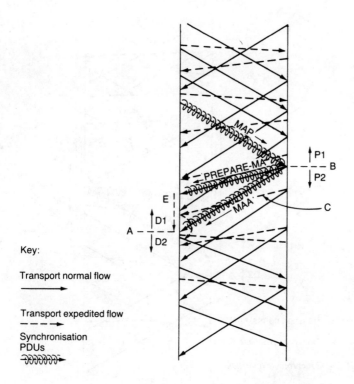

Fig. 6.3 Major synchronisation.

Major synchronisation

First, let us look at the procedures with the help of Fig. 6.3. The initiator of the major synch sends the MAP PDU, and ceases transmission until the synch point procedures are completed.

The responder returns a 'PREPARE' PDU on the Transport expedited flow, and an MAA PDU on the normal flow. This establishes the synch point (point B) at his end.

At the initiation end, all PDUs produce service primitives (part of dialogue D1) *up to the arrival of the 'PREPARE' PDU.* Any PDUs arriving on the *normal flow* after this (area E) are delivered as service primitives in dialogue D1, but PDUs on the *Transport expedited path* are *queued up* (see, for example message 'C' in the diagram), and delivered as service primitives (in dialogue D2) immediately following delivery of the MAA PDU which marks point A, separating dialogues D1 and D2 at the initiation end.

If we now inspect the diagram carefully, we will see that all events prior to A (in both directions) have corresponding events prior to B, and similarly for after. We have a clear dialogue separation.

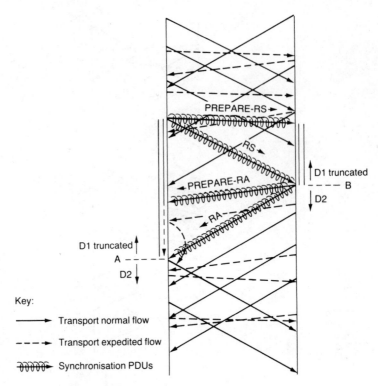

Key:

⟶ Transport normal flow

⟶ Transport expedited flow

⟳⟳⟳⟳ Synchronisation PDUs

Fig. 6.4 Resynchronisation.

Resynchronisation
Again, let us examine the procedure using Fig. 6.4. The initiator of the resynch sends a 'PREPARE-RS' on the Transport expedited flow, and an RS PDU on the normal flow. It then reads and discards *all* incoming PDUs (and ceases to transmit) until it gets a 'PREPARE-RA'.

The responder starts to read and discard PDUs on the normal flow when it gets the 'PREPARE-RS', until it receives the RS PDU. It discards any service primitives issued in this period. (Note: this is not quite true. The user may himself generate an 'S-RESYNCHRONISE' request, and we have *resynchronisation collision*. This adds further complications which are beyond the scope of this text.)

Receipt of the RS PDU marks the dialogue separation point B, and a PREPARE-RA and RA PDU are returned, completing the procedures at this end.

When the initiator receives the PREPARE-RA he *continues to discard normal flow data*, but starts *queueing* (as for major synchronisation) PDUs on the Transport expedited flow. When the RA arrives, this marks the dialogue operation point A; purging ceases, and queued PDUs are delivered as service primitives.

Now let us examine the effects. We have a clean start to dialogue D2. All events *after* A have corresponding events *after* B, and vice versa. Dialogue D1, however, has been truncated. Some events above A have no corresponding events above B, and some above B have no corresponding events above A. In general, neither side knows how much was lost. The diagram shows only the normal data flow of the left to right part of dialogue D1 being truncated. In fact, the expedited part of dialogue D1 can also be truncated in both directions. This is because the Session protocol machine gives 'head of queue' treatment to a PREPARE-RS if it is sitting in buffers behind an Expedited data transmission.

Note also that the 'purging' portions of the procedure (read and discard PDUs) enable resynchronisation messages to get through even if normal flow is blocked, and if the Session expedited flow is blocked, provided this blockage has not produced a blockage of the Transport expedited flow.

A 'government health warning' is being added to the Session standard warning users that too much use of Session expedited can prevent reynchronisation. In this case, a S-U-ABORT and an S-RELEASE will also be blocked. The procedures for S-U-ABORT require the sender of the AB PDU (on the expedited flow in the first version of the Session standards) to set a timer. If the abort is not acknowledged before the timer expires, a T-DISCONNECT is issued. The value of the timer is implementation-dependent.

The detailed description of the use of synch points and their restrictions is contained below.

* Minor synch points: the sender must possess the synch minor token and, if it is in use, the data token also.

The S-Synch-Minor primitive carries a 'type' parameter whose value is determined by the sending Session service user and which is passed to the receiving service user. This allows the sender to request an explicit confirmation of receipt from the receiver. However, the Session service provider does not police the confirmation (i.e. will not detect if it has not been returned). Also, the receiver may confirm receipt even when this was not requested. Note that resynchronisation can be attempted to an unconfirmed synch-point.

The S-Synch-Minor primitive also contains a serial number. The serial number used first was negotiated at connection establishment time. Each S-Synch-Minor primitive carries a serial number one higher than that in the previous primitive (which may have been issued by either service user). The use of a single sequence of serial numbers is possible because the primitive is token controlled. Note that the 'symmetric synchronisation' addendum to the Session service and protocol under preparation by ISO as DAD1 to ISO 8326 and to ISO 8327) provides for minor synch points which are not token-controlled and which use an independent sequence of serial numbers at each end of the connection.

The S-Synch-Minor may also carry an unlimited amount (originally up to 512 octets) of user data.

The S-Synch-Minor response also carries a serial number parameter and an unlimited amount of user data. Because response is optional, there may have been many minor synch points received before any response is issued. The serial number in the response confirms receipt of all synch points up to and including the one bearing that serial number. Note that when an S-Synch-Major response is issued, it automatically confirms receipt of all preceding minor and major synch points (see below).

* Major synch points: the sender must possess the major/activity token and, if they are available, the synch-minor and data tokens as well. This ensures that the receiver will not be sending data or synch-minor primitives when the major synch point is sent. This is, in fact, part of the control of dialogue as described above so that both ends have a common view of the state of the connection.

Because of this property of strong separation between elements of a dialogue before and after major synch points, the term 'dialogue unit' is used to describe the exchanges between adjacent pairs of major sych points.

The only primitives that the sender of the S-Synch-Major can send before receipt of the confirmation are:

(a) S-Token-Give

(b) S-Activity-Interrupt

(c) S-Activity-Discard

(d) S-U-Abort

(e) S-Resynchronise

The receiver of the S-Synch-Major must not initiate:

(a) S-Synch-Major

(b) S-Synch-Minor

(c) S-Activity-Interrupt

(d) S-Activity-Discard

(e) S-Activity-End

(f) S-Release

until the S-Synch-Major response has been issued.

The S-Synch-Major primitives carry two parameters – the synch point serial number and user data.

The serial numbers used in the S-Synch-Major are drawn from the same sequence as those for the S-Synch-Minor. Thus, a synch point serial number uniquely identifies a synch point, major or minor.

An S-Synch-Major response must be issued on receipt of the S-Synch-Major indication. The serial number in the response is thus always identical to the last serial number received in an indication and therefore automatically confirms receipt of all synch points, major and minor, received so far. It also puts all previous synch points 'out of bounds' for resynchronisation using the 'restart' type (see below) of resynchronisation.

Resynchronisation occurs when a Session service user issues an S-Resynchronise request. As both Service users must be aware of the resynchronisation and the resulting state of the connection, this primitive must be confirmed. Thus, when the S-Resynchronise indication is received by the receiving service user, an S-Resynchronise response must be issued and this is delivered to the originator of the request as an S-Resynchronise confirm.

There are three types of resynchronisation, determined by the value of the 'type' parameter inserted into the S-Resynchronise request by the requesting Session service user. In each case, the resynchronise 'purges' the connection (both directions) of any undelivered data or other primitives, leaving the connection in a state where the two users can start again. The primitives also allow negotiation of the location of tokens in a similar manner to that used at connection establishment time.

Most frequently, resynchronisation is likely to take the form of resetting the connection to the last synch point received. The use of 'type' parameter value 'restart' selects this option. Resynchronisation cannot take place to any synch point preceding the last confirmed major synch point but can take place to the last confirmed major synch point or to any subsequent minor synch point.

Resynchronisation may also take the form of 'abandoning' the current dialogue and restarting from a new, as yet unused, synch point. This is similar to the use of the 'break' key on a terminal, for example. Use of the 'type' parameter value 'abandon' selects this option. In this case, the Session Layer itself selects the new synch point number to synchronise to and unacknowledged minor synch points, if any, remain unacknowledged.

The third form of resynchronise, selected by 'type' parameter value 'set', is a combination of the two previously described forms. It allows resynchronisation to any synch point serial number. Note that the restriction to serial numbers subsequent to the last confirmed major synch point does not apply in this case. Note also that this option permits reuse of synch point numbers on a connection.

6.2.2.5 Exception reporting

The Session service provides services for reporting and recovering from error conditions, both within the service provider itself and in the Session service users.

When an error has been signalled, the error condition must be cleared by the service user performing one of the following:

(a) initiating resynchronisation;

(b) initiating an abort 'see below';

(c) initiating interrupt or discard of the current activity;

(d) giving the data token. This last option is really only included to accommodate the CCITT Teletext standard. When used, synch point serial numbers may be lost and any Expedited data sent after the S-Token-Give primitive but which overtake it will be discarded.

Two primitives are used for error reporting as follows.

* S-P-Exception-Report allows the Session service provider to report to both users that an error has occurred in or below the Session Layer. The service provider passes an S-P-Exception-Report indication to both users. When a user receives such an indication, subsequent primitives are discarded until the error condition is cleared by the means described above.

* S-U-Exception-Report is used to allow a service user to report an error. The reporting user issues an S-U-Exception-Report request and this is passed to the receiver as a corresponding indication. The receiver must clear the error condition by one of the means described above. Until this is done, the initiator is not allowed to issue any service requests except S-U-Abort and the service provider will not give data, typed data and Expedited data to the initiator. This service is not available unless the half duplex functional unit has been selected and the data token is not possessed by the initiator.

6.2.2.6 Activity management

The activity management services were introduced into the ISO Session Layer mainly for CCITT compatibility. However, ISO has modified the activity concept in the following way. CCITT activities, as specified in the Teletex standards, are such that normal data, Expedited data and typed data cannot be exchanged outside of an activity. ISO has not made this restriction and as a result also permits use of S-Resynchronise outside activities. Because of this extra freedom in ISO and because the S-Activity-Start is not a confirmed service, there can be cases where the two users have different views as to where in the dialogue the activity actually commenced, particularly in the case where the full duplex funcional unit is in use.

In order to be compatible with the CCITT Teletex standard, an Application standard must forbid use of all forms of data primitive outside of an activity *except* for the capability data (which can *only* be used outside of an activity).

The S-Capability-Data service was conceived by CCITT solely as a means for users to exchange information about their 'capability' to participate in a new activity; hence, the restrictions on its use, i.e. only when the activity management functional unit has been selected and then only outside of any activity.

There are five activity management primitives:

(a) S-Activity-Start: to mark the commencement of an activity;

(b) S-Activity-End: to mark the end of an activity;

(c) S-Activity-Discard: to perform functions similar to resynchronisation;

(d) S-Activity-Interrupt: to break temporarily an activity;

(e) S-Activity-Resume: to restart an interrupted activity.

The primitives will each be described in more detail below:

* S-Activity-Start: a Session service user initiates an activity by issuing an S-Activity-Start request. This is passed to the receiving Session service user as an S-Activity-Start indication. The primitive can only be issued if the major/activity is in the possession of the requestor and the data and synch minor tokens, if available, are also in the possession of the requestor.

The primitives carry an activity identifier parameter which is up to 6 octets long and is chosen by the initiating service user. It is passed transparently to the receiving service user.

The primitives may also carry an unlimited amount of user data.

When issued, the value of the next synch point serial number to be used is set to one, thus preventing resynchronisation to any synch point prior to the start of the activity.

* S-Activity-Interrupt: this is initiated when a Session service user issues an S-Activity-Interrupt request which is passed to the receiver as an S-Activity-Interrupt indication. The sender must possess the major/activity token and cannot issue any further primitives, other than a S-U-Abort, until the corresponding S-Activity-Interrupt confirm has been received.

The receiver of the indication cannot issue any primitives other than an S-U-Abort until it has sent an S-Activity-Interrupt response.

The activity interrupt may cause loss of as yet undelivered data so the users will normally perform resynchronisation when the activity is resumed.

The request and indication carry only one parameter whose value indicates the reason for the interruption. The permitted reason codes indicate various forms of error, a demand for the data token or that the request may not be able to handle further data correctly.

* S-Activity-Resume: a Session service user initiates resumption of an activity by issuing an S-Activity-Resume request which is passed to the receiver as an S-Activity-Resume indication. The request can only be issued if the major/activity token is in the possession of the requestor and, in addition, the data and synch minor tokens, if available, are also possessed by the requestor.

The primitives must carry three parameters:

(a) The identifier for the resumed activity;

(b) The identifier of the previously interrupted activity;

(c) A synch point serial number which is one less than the next number to be used on the connection.

If the activity is resumed on a different connection to that on which it was started, the primitives must also contain the identifier of that previous session connection.

Finally, the primitives may contain an unlimited amount of user data.

* S-Activity-Discard: a Session Service user initiates this service by issuing an S-Activity-Discard request which is passed to the receiver as an S-Activity-Discard indication. An activity must be in progress and the requestor must possess the major/activity token and cannot issue any primitives other than S-U-Abort until the corresponding S-Activity-Discard confirm is received.

The receiver cannot issue any primitive other than S-U-Abort until it has issued an S-Activity-Discard response.

The discard operation terminates the current activity, as does the activity end-service. There is, however, an implied meaning to the service users that the activity has, indeed, been discarded, i.e. all information exchanged since the activity started is discarded and not acted upon (although the Session service provider cannot, of course, check that the service users have acted in this way).

The request and indication primitives carry a parameter whose value indicates the reason for the activity discard. The defined values for this parameter are the same as for the activity interrupt, i.e. errors or inability of the requestor to handle received data correctly.

* S-Activity-End: this is the service used to bring an activity to a normal end. A Service user issues an S-Activity-End request which is passed to the receiver as an S-Activity-End indication. The requestor cannot initiate any services, other than activity interrupt, discard, S-U-Abort or S-Token-Give until the corresponding confirm is received.

   The receiver cannot initiate minor or major synch points, activity interrupt, discard or end, or S-Release until it has issued an S-Activity-End response.

   The S-Activity-End request and indication primitives carry a synch point serial number and convey all the meaning of a major synch point. They may also carry up to 512 octets of user data.

The response and confirm primitives may carry an unlimited amount of user data.

6.2.3 Session connection Release

The Session service provides the means to release the connection in an orderly manner, i.e. without data loss and only by consent of both users. The Session service also provides for connection abort by one of the service users or by the provider itself. Abort destroys undelivered primitives.

The primitives are described in detail below.

* S-Release: this is the orderly release service. A Service user initiates release by issuing an S-Release request which is passed to the receiver as an S-Release indication. The request and indication may carry up to 512 octets of user data.

   The receiver must then issue an S-Release response which is passed to the initiator as an S-Release confirm. The response and confirm primitives may carry up to 512 octets of user data and also carry a result parameter.

   The result parameter may indicate either affirmative or negative. If affirmative, the connection is now closed. If negative, the connection remains open. However, the negative result can only be used if the release token is available, i.e. the negotiated release functional unit is in use.

* S-U-Abort: this primitive allows a service user to abandon the connection in a destructive and immediate fashion. The initiator issues an S-U-Abort request which is passed to the receiver as an S-U-Abort indication. The connection is now closed.

The request and indication may carry up to 9 octets of user data.

* S-P-Abort: this primitive allows the service provider to abandon the connection in a destructive and immediate fashion. The service provider passes an S-P-Abort primitive to both service users. These carry a reason parameter whose value can indicate the reason for the abort to be transport disconnect, protocol error or undefined.

6.2.4 Collisions

It is possible for collisions of primitives to occur. A collision occurs when both Session service users, or one user and the service provider, issue 'destructive' primitives at the same time. A collision is detected when a user or provider which is waiting for a confirmation of a destructive primitive actually receives an indication of another destructive primitive. Neither the users nor the provider can continue with two such operations simultaneously so the service definition defines which primitive should take precedence and which should be dropped in the event of every conceivable collision. Note that the primitive which is dropped may be carrying user data which will therefore be destroyed and not delivered. The rules for handling collisions are provided so that the provider and both users all have a common view of the state of the connection at all times.

6.3   THE SESSION PROTOCOL

6.3.1 Introduction

The Session Layer protocol is specified in the standard ISO 8327. This achieved acceptance as being suitable to be a full IS in late 1984 and is in full alignment with CCITT X.225.

One curious feature of the session protocol standard is the terminology used. Elsewhere in OSI standards the term 'entity' or 'implementation' is used to describe the object which is realising the procedures of the standard. The ISO 8327 uses instead the term 'Session Protocol Machine' (SPM). This, and the style of the document, are the result of having styled the text very much on the state tables, which in this standard take precedence over the text in the event of a discrepancy.

The Session Layer is provided with communications facilities by the Transport Layer which enable it to establish connections of an appropriate quality for the application and to transfer data bi-directionally on those connections. Thus, there should be no need for any protocol mechanisms in the Session Layer other than those directly needed to provide the Session Layer services as described above. None of the error detection and error correction mechanisms normally associated with communications protocols are needed.

However, there are three mechanisms in the Session protocol which can be termed 'optimisation' of the communications channel, even though the Reference Model states that the Session Layer is relieved by the Transport and lower Layers from all concern as to optimisation of the underlying communication medium. This may be due to the limitations of the current transport protocol. In particular, although the Transport protocol has the capability to concatenate TPDUs to form larger NSDUs for efficiency of transmission, it cannot concatenate two data TPDUs together. As Session Layer SPDUs are mostly transmitted in data TPDUs, there may be a fair number of data TPDUs to transmit which are small in size.

Thus, it makes sense for the Session protocol to concatenate SPDUs where possible and this mechanism has been provided.

In addition, the mechanism of segmentation of SSDUs into a number of TSDUs has been provided. The reason for this is in the author's opinion less clear and in fact there has been a recent proposal to have this facility removed from the Session protocol. There is a need for an end-system to be able to control the size of data unit that it is required to handle as has been discussed in the previous chapter. What is not clear is why an end system should be capable of handling large SSDUs but not capable of handling equally large TSDUs since the resources needed are the same in each case.

Finally, the Session protocol permits the re-use of a transport connection for a subsequent Session connection after a previous Session connection has been released. Again, this optimisation has no clear justification, since the time and resource consuming operations in the establishment of an end-to-end connection are located in or below the Network Layer and the Transport Layer is responsible for the retention/release of network connections according to locally defined criteria for the end-system.

In the following sections the protocol will not be described in great detail. This is because to a large extent there is an exact equivalence between the protocol exchanges, parameters and restrictions and the corresponding services. Thus, in general, only protocol mechanisms additional to and not deductible from the service definition will be covered.

6.3.2 Use of the Transport service

The protocol states how the underlying Transport services should be used. There is a general statement and, in addition, a statement for each SPDU as to how that SPDU is to be sent.

* Use of the Transport connection: there are restrictions on how a TC should be (re)used. As there is no multiplexing in the Session Layer, it is necessary to ensure that only one Session connection at a time should be allocated to the TC. In order to avoid the situation where two Session entities decide more or less simultaneously to initiate the establishment of a Session connection, the sending of a Connect SPDU is restricted to the Session entity that initiated the establishment of the TC.

The re-use of a TC is only permitted when the Expedited data service is *not* available on the TC. Even the retention for re-use has to be agreed during the protocol exchanges involved in releasing the previous Session connection. When re-use of the TC has been agreed, the sender of the Disconnect, Refuse or Abort SPDU starts a timer, TIM, and may disconnect the TC if no Connect SPDU has been received by the time it expires.

* Concatenation: there are also restrictions on the way that SPDUs may be concatenated. SPDUs are divided into three categories:

(a) category-0 SPDUs, i.e. those which may or may not be concatenated (Give and Please Token SPDUs). These always occur first in a TSDU containing concatenated SPDUs and there must not, therefore, be more than one in a TSDU;

(b) category-1 SPDUs, i.e. those which are never concatenated (SPDUs concerned with establishment or release of a Session connection, Give Token Confirm and AK SPDUs, and Expedited, Prepare and Typed Data SPDUs);

(c) category-2 SPDUs, i.e. the remainder of SPDUs, which are always concatenated and which always follow just one of the category 0 SPDUs in a TSDU.

There is a protocol option negotiated at Session connection establishment time – the extended concatenation option. If it is agreed this is to be used, more than one category 2 SPDU may be concatenated after one category-0 SPDU.

* Segmentation: the use of the segmentation mechanism, i.e. the ability to subdivide an SSDU into several SPDUs, is implicitly agreed when a maximum TSDU size is negotiated and agreed at Session connection establishment time. When an SSDU is too large to fit within the maximum agreed TSDU size, it must be segmented. Each segment is sent in a separate data TSDU with the last segment containing an 'end of SSDU' marker.

* Transport Expedited: the Transport Expedited data transfer service may be used to transmit some SPDUs. However, the Transport Expedited service is an optional service. The Session protocol specification states under what circumstances the Expedited Service should be requested from the Transport Layer, i.e:

(a) if the Session Expedited functional unit has been agreed;

(b) if the extended control QOS has been agreed.

The Transport Expedited service is used for two distinct purposes. The first is to convey in an Expedited manner an Expedited SSDU. The second is to enable certain services such as abort and resynchronise to be effected efficiently even when the normal data flow channel of the Transport connection is blocked due to congestion.

During the development of the Session Layer standards, there was much discussion as to whether a 'purge' function was needed from the Transport Layer to handle situations where the Transport connection is blocked by flow control, perhaps as a result of one of the end-systems 'hanging up'. In the end it was agreed that the Expedited service could be used to permit a more sophisticated kind of purging.

The basic mechanism is to send a Prepare SPDU using Transport Expedited. The Prepare contains a parameter indicating what the receiver is to prepare for. A control SPDU is simultaneously inserted on to the normal data stream of the Transport connection. The receiver of the Prepare will then start to accept and selectively discard SPDUs from the Transport connection until the awaited SPDU is received. This means that a purging effect can be achieved which, for example, does not destroy synch points.

* Flow control: the Session protocol does not directly provide flow control and instead relies on back-pressure flow control from the Transport Layer. This means that the Session entity will continue to send SPDUs until the Transport Layer indicates that it can accept no more. The way in which this back-pressure is applied is a local matter and depends on the choices made by the implementation.

6.2.3 Connection Establishment

The establishment of a Session follows the sequence described in the service. Here are outlined some of the additional supporting protocol mechanisms.

A Session entity, the initiator, sends a Connect SPDU which is responded to with an Accept or Refuse SPDU. The parameters of the SPDUs are as specified for the S-Connect primitives with the following differences:

(a) there is a protocol options parameter which specifies whether the extended concatenation option can be supported or not;

(b) there is a TSDU maximum-size parameter which specifies, as its name suggests, the maximum size of TSDU that can be handled. It also implicitly requests use of the segmentation option as described above;

(c) there is a protocol-version number parameter which is provided to facilitate later enhancements to the protocol via amendments or enhancements to the standard.

The general style of negotiation adopted in the Session protocol is different to that used in the Transport protocol. In both cases, the initiator makes proposals for the use of options/optional values. However, in the Session protocol the responder then makes counter proposals which are in no way constrained or conditional upon the initiator's proposals. The negotiation is completed when the initiator receives the response with the counter-proposals and both entities use an identical algorithm to determine the outcome based on common knowledge of proposal and counter-proposal.

6.3.4 Data transfer

During the data transfer phase, SPDUs are sent using normal data TSDUs, with the exception of those listed above which are sent using the Expedited data TSDUs.

* Concatenation: When concatenation is used, the receiver is required to process received SPDUs in a predefined order, not the order in which they occur within the TSDU. The processing order is:

(a) 1st – activity start or resume;

(b) 2nd – data transfer;

(c) 3rd – synch points and activity end and acknowledgements thereof;

(d) 4th – token give or please.

* Segmenting: when segmenting is being used, the recipient does not deliver received data or typed data to the service user until the final TSDU of the sequence, indicated by an end-of-SSDU mark, has been received. Receipt of 'destructive' SPDUs before the final segment of data has been received causes the data received so far and not delivered to the service user to be discarded.

* Resynchronisation: the protocol specifies how to mechanise the resynchronise service. Two points in particular are worth noting. The first is the observation that the resynchronise SPDU can be used as a means to forcibly reposition tokens. The second is the definition of how to proceed when two resynchronise SPDUs 'collide', i.e. one is received at a time when another has been sent and not acknowledged. A table is given in ISO 8327 stating precisely which resynchronise SPDU takes precedence and is acted upon (the other being discarded) when a collision does occur. The general principles embodied in this table are:

(a) where the type of each resynchronise is different, the more destructive one takes precedence. Thus 'abandon' takes precedence over 'set' which takes precedence over 'restart' and if both SPDUs are 'restarts' then the one with the lowest serial number takes precedence. Activity interrupt and discard take precedence over all of these;

(b) where the type of resynchronise is the same in each SPDU and the serial number is the same, then the SPDU sent by the initiator of the Session connection takes precedence.

CHAPTER 7

Overview of Presentation and Application Layers

In the preceding chapters we have presented OSI in what may be termed a 'bottom-up' fashion. That is, we have started at the very lowest layer and worked progressively upwards. This has been the most logical order since each layer uses and builds upon the services of the underlying layer. At this point, however, it is appropriate to take a rather different approach since the Application Layer does not 'sit on top of' the Presentation Layer in quite the same way as occurs in lower layers.

First, it will be necessary to introduce some of the concepts of the upper layers and discuss distributed applications.

An application in the sense that is used in OSI is an application to which computer equipment and programs are put. Examples that may be familiar are:

(a) automation of a company's payroll procedures;

(b) automation of processing of cheques through a clearing bank;

(c) holding a database of customer orders in a manufacturing company.

Many applications require the co-operation of a number of computers which are in separated locations, often large distances apart but sometimes only in the next building or floor. These applications are called 'distributed applications'. It is distributed applications that give rise to the need for communications between computers.

OSI does not standardise applications. Instead, the OSI approach, as has been described in Chapter 1, is to standardise the interactions that can occur between those parts of applications that reside in different end-systems (computers). The part of an application that resides in a single end-system is called an 'Application process'. A great deal of what an Application process does is not directly visible from the standpoint of the interactions that occur with another Application process. For example, the actions performed on my behalf by a computer in order that I can create a file of information and subsequently modify it are not reflected in any way in the interactions that take place on a communications link when the file is sent to another computer. The part of an Application process that takes part in the interaction with another Application process is called an 'Application entity'. OSI standards are only concerned with the functions of Application entities, even though the Application entity may be nearly impossible to isolate within a real open system because the code performing those functions can be tightly integrated into the rest of the Application process code.

When the OSI Reference Model was first conceived, a very simplistic view of distributed applications was adopted. It can be summarised as: 'Application process A decides that it needs to communicate with Application process B. As a result A initiates a connection to B, exchanges data with B using the connection and then terminates the connection.'

Now a more complex view of distributed applications is accommodated. For a start, there may be more than two application processes involved in a distributed application and they may interact using a number of two-party dialogues which may or may not be concurrent.

It is also recognised that the nature of the dialogue may change frequently during the life of a single connection. For example, I can connect my microprocessor to a mainframe computer and start to use it as an interactive terminal in order to 'log on' to the computer. I may then choose to transfer an entire file which I had previously prepared on the microprocessor. At that point, I will need to change from interactive (virtual terminal) protocol to autonomous file transfer protocol.

Thus, it is now recognised that a connection may be used sequentially for a variety of Application protocols. Each Application protocol will require its own 'context' which is defined in terms of the Application protocol elements which are active together with a definition of the types of information that may be exchanged, e.g. integers, characters, etc. and described using an 'abstract syntax' such as ASN.1, and the agreed meaning ('semantics') of each type of information, e.g. amount of money to be transferred from one bank account to another.

It is also recognised that elements of Application protocol may be of use to other Application protocols and may therefore be embedded and 'called' upon when required.

In order to describe this internal structuring within the Application layer, the concept of 'Application service elements' is introduced. An Application service element provides some well-defined function. An Application entity may comprise a number of Application Service Elements (ASEs). Two groupings of Application service elements are recognised:

(a) those which will be needed as part of most applications and which are called 'Common Application Service Elements' (CASE). An example of a CASE is the initiation of an Application association;

(b) those which are specific to some style of communication, called 'Specific Application Service Elements' (SASE). An example of an SASE is the Virtual Terminal Service.

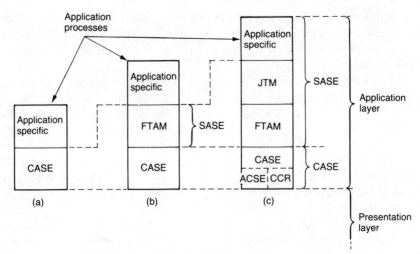

Fig. 7.1   Possible Application Layer structures.

By its nature, a CASE will be called-up from most SASEs. It can also happen that an SASE can be called-up from another SASE (see Fig. 7.1). An example of this might be a banking application calling up a file transfer SASE. This descriptive view of ASEs does not imply any requirement on the structure of an implementation. There is no requirement for CASEs or other ASEs to be written as separate modules of code which can be called within a computer program. An implementation which will only ever operate according to one SASE standard may well choose to embed the CASE elements that are needed into a single piece of integrated code which performs all the Application process functions needed.

The ASEs, and the protocol elements that provide them, make use of communications resources via the Presentation Layer. This enables the abstract data types of the application to be put into a representation for transmission. The definition of a complete set of encoding rules for an application context is called a 'transfer syntax'. Although in concept data is encoded in the Presentation Layer, in practice it may happen that the data is already stored and available in the correct representation for transfer within the end-system and no translation takes place. Thus, the only function of the Presentation Layer which results in externally visible protocol is the negotiation of an agreed set of transfer syntaxes to be used on the connection. Once the use of a Transfer syntax has been agreed, data exchanged between Application entities must be encoded according to that syntax. Within an implementation, this can be achieved in any way that is convenient.

The net effect of these considerations is to view the Application and

Presentation Layers as providing a kit of useful functions and corresponding protocol elements that can be made use of by an Application process according to the rules laid down by the set of standards which define the SASEs which are needed.

Chapters 8 and 9 deal with the Presentation Layer and CASE respectively. The SASE elements currently under standardisation as part of OSI are described in Part 5.

CHAPTER 8

The Presentation Layer

8.1 INTRODUCTION

The Presentation Layer is the sixth layer in the OSI hierarchy. It assumes that an end-to-end path exists, that it possesses the required Quality of Service (QOS); and that any required multiplexing, splitting, segmentation, and dialogue control functions are available from lower layers. It is concerned purely with the representation of data values for transfer.

The Presentation service provides access to all the functionality of Layers 1 to 6. Thus the 'Presentation service provider' is the collection of entities in Layers 1 to 6 of the two end-systems, together with Layers 1 to 3 of all intermediate (relay) systems. The Presentation service users are the Application entities directly using this service.

In distributed information systems, remotely located applications have a requirement to transfer information (semantics), but they are not involved (theoretically) in determining the details of the representation of their information (transfer syntax or encoding) during communication. It is the Presentation Layer of OSI that provides the functions that define, through the use of encoding rules, the transfer syntax for the information being conveyed during the communication.

Application Layer Standards need to specify the transfer of some form of datastructure on a Presentation connection. This datastructure may be a relatively small piece of Application protocol (consisting of perhaps one or more optional parameters, some of which have an internal structure of optional subparameters), or may be a relatively large datastructure consisting of a lengthy report, part of a file or database, or some form of international trade document.

The definition of the meaning to be conveyed (semantics) is recognised in OSI as an Application Layer matter – specific to individual applications.

The conversion of this requirement (to transfer the semantics) into the precise bit-pattern to be carried in Session service user data is recognised in OSI as a Presentation Layer matter.

In general, there are many possible bit-patterns, or *transfer syntaxes*, which (given agreement on the way they are to be used) can be used to convey the same semantics.

At the present time, we have no sufficiently rigorous notation available to enable an Application Standard to specify only the semantics to be conveyed, leaving all aspects of the transfer syntax for determination by algorithms forming part of the Presentation Layer standardisation.

Presently available techniques mean that when an Application Standard specifies the semantics to be transferred, it is forced at the same time to determine *partially* the way that semantics is to be represented. This partial determination of the representation is called an *abstract syntax* definition.

A notation for abstract syntax definition is very similar to the mechanisms in existing programming languages (e.g. Pascal) for defining a datastructure. It provides a set of *primitive elements* whose range of values is fully defined (e.g. Boolean flags, integer values, characters, Application-specific literals) and means of combining them (sequences, choices, sets and iterations). Such a notation enables an Application Standard to formally define the information to be transferred, with minimum constraints on the representation.

The dividing line between what is pure semantics, what is abstract syntax, and what is less abstract syntax(!) is not well-defined. Transfer syntax, however, *is* well-defined. It is the definition of the actual bits to be used to convey the semantics (see Figure 8.1).

Historically, existing Standards, recognised as Application Layer Standards, have chosen to assume their messages were carried by either a binary octet stream or by lines of characters from a fairly basic character set, and

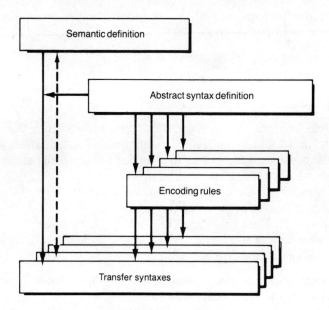

Fig. 8.1 Syntaxes.

have then completely defined their syntax down to this level. Examples are international trade document standards and graphics metafiles.

Such approaches combine together what is now seen as standardisation which is best kept separate. Efficient communication should permit local representation of the necessary datastructures to be used when similar machines are communicating. A clear separation of the transfer from the abstract syntax makes this possible.

Similarly, different transfer syntaxes (encodings, representations) may be required at different times for the same datastructure (abstract syntax). This may be needed to, for example:

(a) use encryption algorithms when needed;

(b) use compression algorithms when needed;

(c) provide verbose but simple and visible encodings when needed.

A separation of encoding issues from semantics, placing the boundary as high as current information technology permits, is clearly desirable. Equally, the ability to handle existing Application Layer definitions which make no such distinction is also essential.

Example:

Suppose we have defined (using an abstract syntax notation) data types called 'Personnel-record', 'Company-report', 'Library-catalogue'. A possible (but perhaps unusual!) value for the user data parameter of a P-Data service primitive would be:

(a) an instance of a 'Personnel-record';

(b) an instance of a 'Company-report';

(c) an instance of a 'Library-catalogue'.

A real implementation represents 'Personnel-records', 'Company-reports', and 'Library-catalogues' in some manner suited to the programming language used for writing the Application protocol handler. (This will typically involve pointers and a tree.)

The issue of a P-Data primitive involves passing a pointer to this data structure to the Presentation Layer module, for it to encode the values and issue an S-Data primitive. (In order to do this, the Presentation Layer module will, in general, also require a pointer to a data structure or file containing the type definitions.)

The Presentation Layer also provides a common service to applications for negotiation and agreement of the 'transfer syntax' for the information the applications wish to communicate. Figure 8.2 illustrates the use of the

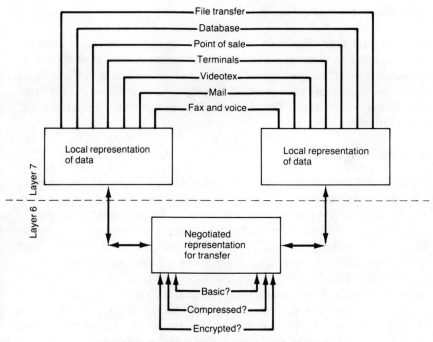

Fig. 8.2 Operation of Layers 6 and 7.

Presentation Layer to negotiate the transfer syntax for a range of applications.

This chapter describes the provision of the Presentation Layer services that facilitate the definition of (structured) data values and the specification of encodings (representations, transfer syntax) for such values. The applicable standards are CCITT Recommendation X.409, and ISO Standards 8822, Connection Orientated Presentation Service definition: 8823, Connection Orientated Presentation Protocol specification: 8824, Specification of Abstract Syntax Notation One (ASN.1); and 8825, Encoding Rules for ASN.1.

8.2 PRESENTATION LAYER CONCEPTS

The Presentation Layer provides for the negotiation and establishment of the transfer syntax, which represents the encoding of values for the purposes of transfer of (structured) data types. A data type may have more than one possible transfer syntax, providing greater or lesser degrees of compression or security.

8.2.1 Abstract syntax definition

A set of data-type definitions, using a well-defined notation, constitutes what is called an abstract syntax for the information that is contained in values of any of the data types. Such a well-defined notation could, in theory, be that of any programming language with a sufficiently rich data-structuring notation (e.g. Pascal, ADA, etc.) or a variant of Bachus-Naur Form (BNF).

8.2.2 Encoding rules

To be useful for Presentation Layer purposes, a notation for defining the abstract syntax of data values needs one or more accompanying sets of encoding rules that determine algorithmically, for any values of any set of data structures defined using that notation, the transfer syntax (representation) to be used. Programming language (and BNF) notations for data structure definitions normally lack such a set of encoding rules.

The CCITT and ISO have defined an abstract syntax notation, called Abstract Syntax Notation One (ASN.1), and a corresponding set of Basic Encoding Rules. The specifications in ISO 8824, ISO 8825, and CCITT Recommendation X.409 form one means of support for Presentation Layer functions. They are described later in this tutorial. The possibility of other notations and encoding rules providing such support is also explicitly recognised by ISO.

ISO recognises that there may be a need for more than one notation for abstract syntax definition. This arises for two reasons:

(a) the desire to move the level of abstraction of such definitions closer and closer to pure semantics;

(b) the recognition that different applications will have very different primitive elements and construction techniques.

The latter point is illustrated by comparing the ASN.1 notation described later with a notation designed to handle graphic information. In the latter case, primitives may be points, colours, shadings, line markings (dotted, dashed, full, etc.), point markings (triangle, cross, square, etc.) and the construction mechanisms may be polygonal forms, arcs, splines, infills, ellipses, and so on, together with overlaying, windowing, and rotations.

Other examples of requirements for differing abstract syntax notations occur in videotex communication and in digital voicework.

ISO has developed a Standard for an Abstract Syntax Notation (ASN.1), and for encoding rules which define the transfer syntax for any datastructure whose abstract syntax is defined using ASN.1. (This work is further described below.)

8.2.3 Presentation context

The Presentation Layer Standards require the allocation of unambiguous names to two sets of objects.

The first requirement is for an unambiguous name for a set of data structures to be used on the connection. These datastructures will typically (but not necessarily) be defined using an abstract syntax notation. This set of datastructures is called *a named abstract syntax.*

The second requirement is for unambiguous names for one or more defined transfer syntaxes capable of carrying the semantics of a particular abstract syntax. These transfer syntaxes will typically (but not necessarily) be defined by referencing the algorithms of some encoding rules and applying them to the abstract syntax definitions.

The Presentation service enables a service user to specify a list of *named abstract syntaxes* for this connection by giving their unambiguous names. The Presentation Layer protocol then negotiates which of the available transfer syntaxes (using their unambiguous names) is to be used for each abstract syntax. The resulting combination of a single abstract syntax and the agreed transfer syntax is called *a Presentation context*; it is given a local identification by the Presentation service for the duration of the connection. (This is illustrated in Fig. 8.3.)

Thus, the Presentation Layer negotiates the transfer syntax to be used for data values associated with an abstract syntax name. This association between an identified set of data values and their representation constitutes what is called a defined Presentation context. Such contexts have a lifetime that is limited to a single Presentation connection, and needs to be redefined if the connection is broken and re-established.

At any moment, several Presentation contexts may have been defined by negotiation of transfer syntaxes. These form what is called the Defined Context Set (DCS).

Example:

Suppose the data type definition 'Personnel-record' is part of a set of data type definitions that are given the abstract syntax name 'Staff-data'. Suppose, further, that a simple transfer syntax for all the data types in 'Staff-data' has been given the transfer syntax name 'Staff-Data-Encodings (BASIC)', and that a second transfer syntax has been given the name 'Staff-Data-Encodings (COMPRESSED)'.

The Presentation Layer might establish, for the duration of a connection over a high-bandwidth line, a Presentation context for the transfer of values from 'Staff-Data' using 'Staff-Data-Encodings (BASIC)'. This line might fail during the (check-pointed) transfer of values. Re-establishment of the Presentation connection might necessi-

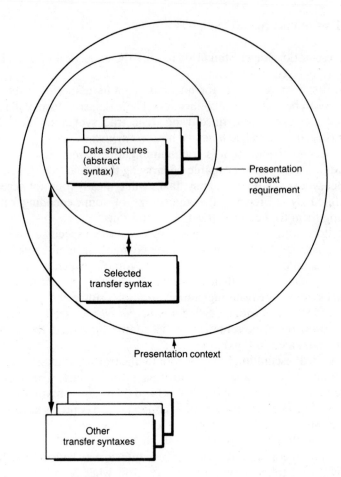

Fig. 8.3 Presentation contexts.

tate the use of a low-bandwidth line. In this case, the Presentation context for continuing the transfer of 'Staff-Data' values from the checkpoint might be established using 'Staff-Data-Encodings (COMPRESSED).

Finally, the Presentation Layer recognises the possibility of identifying use of the Session services 'by prior agreement', and that such 'prior agreement' could involve the definition of a 'default context', and omission of all P-Connect parameters because their values are already known (by the 'prior agreement') to both ends. In this case the P-Connect is null. This is done for compatibility with existing protocols such as the X.400 series, which make direct use of the Session service.

Example:

If a context is defined during a connection to support the values of data types defined in the set with the abstract syntax name 'Staff-Data', a single Presentation context is defined. Another could be defined at the same time or later in order to support data types (including 'Company-report') defined in 'Commercial-Data'. Similarly, a context could be defined to support 'Library-Data'.

These context definitions are remembered by both parties to the communication, and form what is called the *defined context set* (DCS). One or more of these contexts could be later deleted, or another context definition added by negotiation, forming a new DCS.

If the three contexts defined above form the DCS, the Presentation Layer permits the transfer of sequences of values of any data type in 'Staff-Data', 'Commercial-Data', or 'Library-Data', in the same P-Data primitive. It does not permit the transfer of values of any other data type.

It is possible to establish several Presentation contexts for use on a connection. These may involve different transfer syntaxes for the same abstract syntax, or may involve different abstract syntaxes. Note that if a receiver is to decode successfully an incoming message (set of bits) it needs to know the abstract and transfer syntaxes involved, i.e. the Presentation context in which the data is transmitted.

Within a single Presentation context, it is the job of the definer of the abstract syntax and of the encoding rules to ensure that a receiver of data can interpret it unambiguously.

The Presentation Layer undertakes to signal any change of Presentation context. Moreover, it introduces the concept of *multiple active contexts*. This is the case where, at any instant of time, the sender can transmit an instance of a datastructure from more than one context. Potentially, two such different datastructures could coincidentally have the same bit-pattern representation, and hence be ambiguous to the receiver. The Presentation protocol specifies how this potential ambiguity is resolved.

8.2.4 Identifying data values on reception

A specification of an abstract and transfer syntax is not well-formed unless, for each data value associated with the corresponding abstract syntax name, there is a distinct representation that enables a receiver to distinguish that value from all other data values associated with the same abstract syntax name.

The transfer syntax for data values from different Presentation contexts will frequently (by accident) be an identical set of octets. A major function of the Presentation protocol is to ensure that a receiver is aware of the Presentation context in which any received octets are to be interpreted and, hence, can correctly identify the data value.

8.2.5 Presentation service user data

Presentation service user data parameters consist of an indefinite number of data values, with each value associated with the same or different Presentation contexts.

8.2.6 Primitives and parameters

All Presentation primitives correspond to Session Layer primitives of the same name, with the exception of P-Alter-Context which is the only primitive introduced by the Presentation Layer.

Presentation service parameters have the same meaning as in the Session service, with the following additions:

Multiple defined contexts	This parameter is a flag used to negotiate the availability of multiple defined contexts at any one time. Note that if this is not set, the context of user data is always known without tagging.
Presentation context	This is a list of abstract syntax names, to each of which the Presentation protocol entity adds a list of transfer syntax names. Negotiation results either in a defined context for all of the entries in the list, or failure of the P-Connect.
Default context definition (optional)	This is a single abstract syntax name to which the protocol adds a single transfer syntax name. There is no negotiation. The connection is refused if this is not acceptable to the responder.
	The default context, if defined, applies when there are no entries in the DCS. It is also used for

	user data carried by primitives that can overtake primitives recording DCS changes, and that therefore need to be independent of the DCS. (These include P-Expedited-Data, P-U-Abort, and P-U-Exception-Report.)
Presentation requirements	This parameter negotiates the availability of the P-Alter-Context service.
Result	This parameter signals success or failure of the requested service.
Result list	This parameter informs the recipient of a P-Alter-Context indication of any context that the provider cannot support; in the response and confirm it is equal, and ensures that all parties are aware of the resulting additions to the DCS. (The additions occur at the two ends when the response and confirm are issued.)
Restore indication	This is discussed in the resynchronisation section below.

8.2.7 Resynchronisation

The effects of resynchronisation on Presentation Layer functions are a direct consequence of the nature of the Session service and decisions in the Presentation protocol.

In order to define new contexts (add to the defined context set) or delete definitions (remove from the DCS), a Presentation protocol exchange is required to support the confirmed service of P-Alter-Context.

With the current Session service, this exchange can only be safe against disruption by a *later* use of P-Resynchronise (mapping to S-Resynchronise) if it is carried using S-Sync-Major. In this case, context definition would have the side-effect of setting a major synchronisation point and preventing later resynchronisation to an earlier minor synchronisation point. There are cases that can be envisaged where this side-effect of defining a new Presentation context would be highly undesirable. In general, Applications

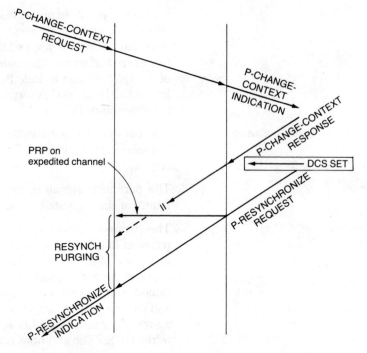

Fig. 8.4 Destruction of DCS consistency by resynchronisation.

need to be able to exercise independently the functions of check-pointing and defining contexts.

An alternative would be for the Session Layer to provide an additional service for the transfer of data, free from disruption by a later S-Resynchronise. This, however, is not present in the Session service.

As a result of these considerations, Presentation protocol data units supporting P-Alter-Context are carried by S-Typed-Data, and are subject to disruption by a later P-Resynchronise as shown in Fig. 8.4. This means that use of a P-Resynchronise request after a P-Alter-Context response (possibly some time after) leaves the Presentation entities with different opinions on the contents of the defined context set. This would result in the possibility of one end sending data that the other end could not interpret. The protocol would not work.

It would be possible to rely on Application Layer protocols to specify 'Application rules' to avoid such collisions, but the Presentation protocol designers decided to, in effect, carry the latest sent P-Alter-Context message in the P-Resynchronisation PDU.

Note also that the use of S-Typed-Data for context definition means that a change to the DCS, which would be a P-Alter-Context response, can also be overtaken by a later use of P-Expedited Data, P-U-Abort and P-U-Exception-Report, again resulting (potentially) in different views of encoding algorithms for the user data in such primitives. This is resolved in a similar way.

The above represents the 'basic' Presentation service, without the 'context restoration' functional unit. Resynchronisation has no effect on Presentation contexts. There are cases, however, where this is not what the Application protocol designer will want. In resynchronising, he expects his DCS to be *restored* to that which obtained at the synchronisation point.

In this service, the Presentation Layer identifies points where it can guarantee that the DCS is well-defined, and 'remembers' the DCS at these points. If resynchronisation occurs to one of these points, the DCS can be restored to the value that has been 'remembered'. Figure 8.5 lists the points at which the DCS is well-defined, and will be 'remembered'. Provision for P-Sync-Minor and P-Activity-Begin is made by forbidding the issue of these requests while context definition is in progress.

Each major or minor synchronisation point issued when the activity functional unit is *not* available has the corresponding DCS 'remembered' for as long as it is current (available for S-Resynchronise (Restarts)).

When activities are in use, each synchronisation point 'memory' is labelled with the Session 'activity identifier' parameter used on the S-Activity-Begin or S-Activity-Resume (if an activity is in progress).

A P-Activity-Discard or P-Activity-End loses such 'memory'. A P-Activity-Interrupt does not. Provided that P-Activity-Resume has the same 'Session activity identifier' and that the current 'Session connection identifier' occurs within this connection, the 'memory' is still available for restoration. (Note, however, that a P-Activity-Resume for the same activity identifier with a *different* 'old Session identifier' *loses* the memory.)

8.3 ABSTRACT SYNTAX NOTATION

At the present time, there is only one notation under development in ISO for defining the abstract syntax of a datastructure to be transferred between two systems. ISO recognises that this may be the first of many and has called it 'ASN.1'.

This notation was originally developed by CCITT in order to define the fairly complex datastructures transferred as part of their message-handling systems (electronic mail).

In this work, the notation is used not only to define the content of protocol elements and 'envelope' information, but also to define the form of 'office documents' – letters, papers, and so on, themselves.

CCITT published the work as Recommendation X.409. This Recom-

P-Connect	response and confirm
P-Sync-Major	response and confirm
P-Sync-Minor	request and indication
P-Activity-Begin	request and indication

Fig. 8.5 Points of a well-defined DCS.

mendation contains, in a single text, both the definition of the notation and the definition of a set of encoding rules which enable a transfer syntax to be derived from any abstract syntax definition using the notation.

ISO has separated the encoding rules from the definition of the notation, producing two Standards which are technically identical to X.409, but completely rewritten. It is expected that in 1988, ISO and CCITT will adopt a common text for the further development of this work based on the ISO Standards.

ISO has identified a number of areas for possible extension of ASN.1, but has not included them in the initial Standards, in the interests of CCITT alignment and co-operation.

ASN.1 recognises the following primitive types (data elements) from which more complex data structures can be assembled:

(a) Boolean, a true or false value;

(b) Integer, the positive and negative whole numbers, including zero; the range is effectively unbounded; particular values can be named;

(c) BitString, an ordered set of zero or more bits of unbounded length; the length need not be known in advance of transferring the initial bits; particular bits can be named;

(d) OctetString, an ordered set of zero or more octets of unbounded length; the length need not be known in advance of transferring the initial octets;

(e) Null, a primitive type consisting of a single value, also called Null; it can be used as a place-holder when optional elements are missing.

Complex datastructures are defined by applying 'constructions' to these primitive types or to types produced by earlier constructions. The main constructions available are:

(a) Sequence, a value of the new type is an ordered list of values of specified existing types;

(b) Sequence of, a value of the new type is zero one or more (unbounded) values of a single existing type;

(c) Set, as for sequence, but order is not significant;

(d) Set of, as for sequence of, but order is not significant;

(e) Choice, a means of specifying that the value at this point in the datastructure can be the value of any of a list of defined types.

In addition, ASN.1 defines *character string types*, such as 'VisibleString' – an ASCII character string, and what it calls 'useful types' such as time and date.

Example:
The following is a fictitious example of the use of ASN.1 (or X.409) to define a data structure. It is provided as the simplest way of getting a 'feel' for the notation. (*Note:* Words beginning with a capital letter are the names of types. Words beginning with a lower-case letter are identifiers for human readability, and do not contribute to bits on the line.)

```
DEFINITION ::=
BEGIN
PDU ::= SEQUENCE
        {fixed-part          SEQUENCE
          {protocol-id          INTEGER,
           called               Address,
           calling              Address}              ,
        variable-part        SET
          {class                 INTEGER
                                 DEFAULT 0 ,
           connection-id         GraphicString
                                 OPTIONAL   ,
           security              CHOICE
             {high                (0) NULL,
              medium              (1) NULL,
              low                 (2) NULL}
              Default             {medium NULL},
           user-data             OCTETSTRING
                                 OPTIONAL,
           options               BITSTRING
             {kernel              (0),
              management          (1),
              access              (2),
              transfer            (3)}
                   DEFAULT '1000' B                   },
        routing-info         SEQUENCE OF
                             Address                   }
```

```
        Address ::=SEQUENCE
           {nsap-address NumericString
            -- up to 40 digits --      ,
            ts-selector OCTETSTRING
            -- up to 8 octets --       ,
            ss-selector   ISO 646STRING      }
   END
```

The PDU-data type definition consists of a collection of fields that, at the lowest level, consist of a human-readable identifier, a possible 'tag' (e.g. '[1]'), a type-reference, and a possible indication that the field can be omitted (OPTIONAL), or that it can be omitted and also has a default value (DEFAULT). (*Note:* in order to support 'DEFAULT', ASN.1 defines a notation for *value* specification as well as for *type* definition.)

These lowest-level fields of the definition are combined using structuring mechanisms that begin with the name of the type of the structure, and then, in general, a list of fields separated by commas and enclosed in curly brackets. The use of structuring mechanisms can be nested to any depth.

The human-readable identifier is important for the value notation, but is ignored by the Basic Encoding Rules. It commences with a lower-case letter, and consists of letters and hyphens.

The 'tag' enables two apparently identical types to be separated so that their use can be distinguished in constructions such as Choice or Set (where the order of transmission is not constrained).

The type-reference is either a type that is defined using the ASN.1 notation (such as 'Address' above), or is a type defined in the ASN.1 standard. These types are discussed below.

The various structuring techniques provide for iterations of a type (Set of and Sequence of), listing of types (Sequence and Set), and specification of a type as a type taken from a set of alternatives (Choice). These techniques are discussed below.

The outermost feature of the notation is the 'module'. A module is a set of type definitions that are complete and consistent. The module name (preceding the word Definitions) can be used to enable modules to reference types defined in a different module. Thus, 'Example.PDU' could be used in another module to reference the type named 'PDU'.

Module names used by ISO are expected to include the number of the standard and its acronym, e.g. the Commitment, Concurrency, and Recovery (CCR) use of ASN.1 is carried in the module named ISO 8650-CCR.

The above definitions would be part of Application standardisation. The ISO Standard defining the encoding rules for ASN.1 ⌐.ın now be applied to determine the precise set of bits which will be sent down the line. The resulting abstract and transfer syntaxes would need unambiguous names for use during communication.

ASN.1 also defines a human-readable notation to be used for writing down values of data types defined using ASN.1.

The combination of the basic ASN.1 structuring mechanisms and a good 'library' of Application-specific types, provides a very powerful tool for designing high-level communications. Moreover, the resulting definition is fairly readable to the non-expert.

ASN.1 provides one further powerful mechanism. This is the so-called 'macro notation'. This permits a designer to extend or change the ASN.1 notation itself. Suppose, for example, in the interests of human-readability you preferred to write

Either Exercise or Operation or Neither

instead of

Choice, {Exercise, Operation} Optional

This alternative notation can be established by use of an ASN.1 macro – but we are on to advanced use now!

Most of the extensions proposed by ISO to ASN.1 relate to the addition of further character string types, new 'useful types', and formalisation of means of indicating restricted ranges for integers, lengths of strings, and number of iterations.

Whilst originally designed by CCITT to cover the requirements of message handling and office document formats, ASN.1 is being used by the ISO Standards for JTM, CCR, FTAM and VTP to define their protocol elements. It appears likely to be useful in a wide area of applications.

The ASN.1 Standard contains a number of examples, in annexes, which further illustrate the use of the notation.

8.3.1 Abstract Syntax Notation One defined types

Some basic data types defined by ASN.1 are:

(a) Boolean

(b) Integer

(c) BitString

(d) OctetString

(e) Null

(Useful with tagging for CHOICES, as in 'security' in the example)

For Integer (and BitString) types certain values (or bit positions) can be given identifiers for use by humans; this is illustrated by 'options' in the example.

There are also 'character string types'. These are:

NumericString	(Digits and spaces)
PrintableString	(Upper and lower-case letter, digits, space, and eleven other characters)
IA5String	('ASCII' characters, and carriage controls)
VisibleString	(Printing characters only from the 'ASCII' character set)
TeletexString	(T61String in X.409)
VideotextString	
GraphicString	(Printing characters from any character set registered in the ISO Register of Character Sets; this includes several Japanese, Chinese, and Arabic character sets, as well as a large number of 'European' versions of ISO 646 (International equivalent of ASCII). *Note:* this type is not currently included in CCITT X.409.)
GeneralString	GraphicString plus control characters.

Finally, the 'useful types' defined in ASN.1 are:

GeneralizedTime
UTCTime

These types provide for various degrees of flexibility and precision in specifying dates and times.

8.3.1 Abstract Syntax Notation One structuring mechanisms

The following structuring mechanisms are available:

(a) Sequence	an ordered list of types;
(b) Sequence of	an (unbounded) iteration of a single type (order significant);

(c) Set   an unordered list of types;

(d) Set of   an (unbounded) iteration of a single type (order not significant);

(e) Choice   a field that consists of a value from any one of several listed types.

8.3.3 The macro notation

ASN.1 defines a means by which a (sophisticated) user can specify a new (and arbitrary) notation for defining ASN.1 types and values of those types (*Note:* although the new notation is arbitrary, the range of possible types is still limited to those that can be described by the ASN.1 structuring mechanisms applied to the simple types.) The major use to date of this macro notation is in the X.410 specification of 'Remote Procedure Calls'.

8.3.4 The 'External' notation

ASN.1 (but not the initial X.409) provides notational support for embedded values from other Presentation contexts, as described earlier. The notation is simply the word 'EXTERNAL'.

8.3.5 Future work

Several extensions to ASN.1 are proposed. Perhaps most noteworthy are the inclusion of floating point numbers and of pointer types to support a general 'mesh' type of data structure.

8.4 BASIC ENCODING RULES

The Basic Encoding Rules provide an algorithm that specifies how a value of any data structure (type) defined using ASN.1 is to be encoded for transmission.

Use of the algorithm in reverse enables any receiver *who has knowledge of the ASN.1 type definition* to decode successfully an incoming bit stream into a value of that type.

The encoding of an ASN.1 data type (using the Basic Encoding Rules) allows a receiver, without knowledge of the type definition, to parse the incoming octet stream to identify the start and end of constructions (Sequence, Set, etc.) and the octets representing basic data types (Boolean,

Integer, etc.). In the simplest use of the notation, it is also possible to determine from the encoding the actual construction and basic data types in use.

When an ASN.1 tag (e.g. the '[1]' in the earlier example of use) is employed, the encoding carries both the value of the tag *and* a value identifying the construction of the basic data type that is being tagged.

It is also possible, in the ASN.1 notation, to use what is called *implicit* tagging. For example:

Example-Type::   = CHOICE

{first-choice   [0] Implicit Integer,

second-choice   [1] Implicit Boolean

In this case, the tag value (zero or one) is present in the encoding, but the codes to identify the basic data type as Integer or Boolean are suppressed. The receiver needs to have available the actual type definition in order to determine the basic data type. This is the general case.

The Basic Encoding Rules have a TLV structure (*T*ype, *L*ength, *V*alue). Each data value to be encoded is given a representation consisting of:

(a) one or more octets identifying the type of the value within some scope; (typically the type will be the 'universal class' type assigned by ASN.1; or a context-dependent type assigned by 'tagging' in the type notation);

(b) one or more octets specifying the length of the value, or specifying an 'indefinite' length;

(c) octets representing the value; for simple values (Boolean, Integer), these octets have no further structure (they are leaves in the parsing tree); for structured values (Sequence, Set, etc.) these octets are themselves a series of TLV components.

The 'indefinite' length indication is only available if the value is structured, i.e. a series of TLV components. The end of the series is indicated by two zero octets, with the additional design feature that a 'T' *never* begins with a zero octet.

The main use of the 'indefinite' form is to permit the transmission of material whose length is unknown when transmission begins. It is clearly important for this to be available for OctetString and BitString. Normally, such types (and types obtained from them by tagging), would be simple, and hence ineligible for the 'indefinite' form. A special mechanism is provided to enable these types to be encoded as either simple or as structured. If they are encoded as structured, each element of the structure is a TLV-encoded OctetString (or BitString respectively), i.e. a fragment of the main string.

8.5 SUMMARY AND CONCLUSIONS

This chapter has described four ISO draft Standards and one CCITT Recommendation. The main features of the Presentation service (ISO 8822) and protocol (ISO 8823) are:

(a) separation of Layer 7 definition of abstract syntax from the Layer 6 concern with encoding;

(b) negotiation of transfer syntaxes;

(c) stacking, destruction, and restoration of the defined context set (DCS);

(d) identification of the context of a data value, delimiting a transfer syntax, and octet alignment;

(e) null protocol in cases of extreme optimisation and 'prior agreement'.

The main features of ASN.1 (ISO 8824) and its Basic Encoding Rules (ISO 8825) are:

(a) a general notation for abstract syntax definition;

(b) a range of basic data types;

(c) a range of 'character string types';

(d) other useful types;

(e) a TLV type of encoding;

(f) support for character, octet, and bit strings whose lengths are not known in advance of the initial transmission;

(g) CCITT X.409 compatibility.

These four Standards form a major part of the OSI suite. They not only establish a general-purpose Presentation Service and protocol in a form independent of any syntax notation, but also provide initial standards for a highly useful notation and for encoding rules.

The ASN.1 notation is being used by all Application Layer protocols currently under development by ISO and CCITT. The framework is well established for future development, and the need for compatibility with existing 'direct' approaches has been recognised. Although CCITT involvement in the main Presentation protocol was minimal, its contribution to the ASN.1 work has been dominant.

CHAPTER 9

Common Application Service Elements

9.1 Introduction

Chapter 7 introduced the concept of Common Application Service Elements (CASE) and explained that they represented application service elements (ASE) which are needed by a large number of Application entities. Currently there are only two CASEs:

(a) association control, providing service elements for the establishment and termination of an application association;

(b) commitment, concurrency and recovery (CCR), providing service elements for the co-ordination of activities across several Application processes such that it is always possible to recover to a well-defined position following failures in communications resources or end systems.

It is possible that in time the introduction of more SASEs will cause recognition of the need for more CASEs. The need for a CASE to control the swapping of application contexts on an association is already under discussion. This is needed in order to support multiple concurrent associations on a single Presentation connection and in order to support sequential re-use of a Presentation connection by Application associations.

Each SASE standard will specify which CASE it makes use of and how the SASEs are carried in, or otherwise make use of, each CASE.

CASE is defined by two multi-part standards:

(a) ISO 8649 which defines the services provided;

(b) ISO 8650 which defines the corresponding protocol.

Both are in three parts:

(a) Part 1 is simply an introduction which describes the further parts of the standard and which will be updated whenever a new CASE is standardised;

(b) Part 2 defines association control;

(c) Part 3 defines CCR.

One problem that has arisen for both the Application and Presentation Layers has been that of alignment with and accommodation of the CCITT X.400 message-handling recommendations. The X.400 recommendations were developed by CCITT in order to provide the basis for future electronic mail services and were first published in 1984. At that time the services and protocols for the Application and Presentation Layers of OSI were not very advanced. CCITT therefore built their X.400 protocols to operate using only the services presented by the Session Layer.

In order to allow the X.400 protocols to be supported by the full services of OSI, ISO is about to introduce (at the time of writing) 'transparent' modes for the Presentation and CASE association control protocols. In both layers it is permitted to proceed to the data phase without protocol exchanges for connection/association establishment in cases where the Presentation and Application contexts have been established by 'pre-arrangement'.

The remainder of this chapter will define the association control and CCR service elements and corresponding protocols.

9.2 ASSOCIATION CONTROL SERVICE ELEMENTS (ACSE)

The Association Control Service Elements (ACSE) provide the following services to SASEs:

(a) A-Associate which enables an Application entity to establish an association with another Application entity. The entities may identify themselves by use of their titles and may specify an application context that applies to the association.

(b) A-Release which allows for the orderly, negotiated release of the association without loss of information.

(c) A-Abort which allows for an Application entity to cause an immediate and possibly disruptive release of the association.

(d) A-P-Abort which allows the service provider to cause an immediate and possibly disruptive release of the association and inform both Application entities.

At the time of writing the ISO 8649 only allows for a one-to-one correspondence between an Application association and a Presentation connection. It is possible that later extensions will permit a Presentation connection to be re-used for a new association after an existing association has terminated or to permit multiple associations to be interleaved on to a single Presentation connection. If a pair of end-systems wish to support several independent Application associations simultaneously then it is always possible within the existing standards by using the multiplexing function of the Transport Layer.

9.2.1 Service elements

9.2.1.1 A-Associate

The A-Associate is a confirmed service and the primitive therefore exists in request, indication, response and confirm forms.

The following parameters may be included with each form of the A-Associate as specified below. For each CASE-specific parameter (as opposed to those which are Presentation P-Connect parameters) SASE must specify requirements and conditions for use. CASE defines the CASE-specific parameters to be optional. The CASE-specific parameters are:

(a) Recipient (destination) Application entity title. The inclusion of this parameter (in the request) is optional. When included it will be present in the indication with the same value as in the request. The parameter value in the response need not be the same as in the Request and Indication. This allows for the concept of generic titles whereby the title of a *type* of Application entity (e.g. a type which provides a particular service) can be requested and the response may be returned from a specific invocation of the Application entity type.

(b) Initiator Application entity title. The presence of this parameter is optional and it is included in request and indication only.

(c) Application context name. This parameter allows the Application context requirements for the initiating SASE to be specified. The value in the indication will be the same as the value in the request when included. The value in the response need not be the same as the value in the indication. This allows for the possibility of a limited amount of context negotiation. The semantics of such negotiation are, however, undefined and must be specified in each SASE standard. It could be used, for example, to negotiate options within the Application context or a version of the Application protocol. When the Application context parameter is not supplied, it is assumed that both Application entities have agreed an Application context by prior agreement.

(d) User information. The meaning of this parameter, if supplied, is dependent on the Application context.

(e) Result. This parameter is only available on the response and indication primitives. The parameter indicates whether the request was accepted or rejected and, if rejected, the reason for rejection.

All the Presentation service P-Connect parameters are also available in the A-Associate. Their possible values, semantics and usage are defined in the Presentation service standard ISO 8822. Each SASE will specify its Presentation service requirements.

If the SASE specifies, by means of the Presentation service parameters, the use of a single Presentation context then the ACSE abstract syntax must form part of the specified Presentation context.

If the SASE does not specify the use of a single Presentation context the multiple contexts facility will be used and the ACSE service provider will exchange A-Associate semantics using its own dedicated Presentation context.

The Presentation service parameters also select Session services and the choice of Session services affects other CASE operations as follows:

(a) if the use of tokens is implied by the choice of Session functional units then the A-Release service can only be invoked by the Application entity that holds the tokens;

(b) the recipient of the A-Release indication can only refuse the release request if the negotiated release Session functional unit has been selected.

9.2.1.2 A-Release

The A-Release service is the normal method of terminating an Application association. Use of A-Release brings the association to an orderly end without loss of data in transit and, if the Session Layer negotiated release functional unit has been agreed for use on the connection (via the equivalent Presentation Layer service), allows the responder to reject the A-Release and so preserve the association. A-Release is a confirmed service and so there are the usual four types of A-Release primitive (request, indication, response and confirm). A-Release does not disrupt any other service.

There are three parameters for each of the four A-Release primitives:

(a) Reason: which takes the value 'normal', 'urgent' or 'undefined' in the request and the same value on the indication. The response may take the value 'normal', 'not finished' or 'undefined' and will take the same value in the confirm.

(b) User information: whose use is specified by the SASE corresponding to the current Application context.

(c) Result: which is only carried by the response and the corresponding confirm and which may take the value 'affirmative' or 'negative'.

When the A-Release request has been issued, the ACSE user may not use any further primitives except A-Abort. The ACSE releases the Presentation connection simultaneously with the release of the Application association (in practice this means that the A-Release PDU is embedded in the P-Release service primitive). On receipt of the A-Release indication the responder issues an A-Release response which may carry a 'Result' parameter with the value 'negative' (if negotiated release is available) in which case the association remains open, or which may contain the value 'affirmative' in which case the association is terminated as far as the responder is concerned. The association is terminated for the initiator when the A-Release confirm is received with 'result' equal to 'affirmative'.

Collisions between simultaneous A-Release primitives and between A-Release, A-Abort and A-P-Abort primitives are resolved by the Session Layer procedures described in Chapter 6.

9.2.1.3 A-Abort

The Application association may be forcibly terminated, with possible loss of data in transit, by either Application entity by issuing an A-Abort request.

The remote Application entity is informed and the association is terminated. The Service is unconfirmed. The A-Abort disrupts all other services except the A-P-Abort.

The A-Abort primitive carries two parameters:

(a) Abort source: which may take the value 'ACSE service provider' or 'requestor';

(b) User information: whose meaning depends on the Application context.

9.2.1.4 A-P-Abort

The A-P-Abort service element allows the ACSE Service provider to signal an abnormal termination of the association due to loss of the underlying Presentation connection (which may be due to one of many reasons such as irrecoverable loss of communication in the lower layers or problems in the end-system itself). The A-P-Abort service, if it occurs, will disrupt any other services in progress.

The parameters of the A-P-Abort are those occurring in the P-P-Abort service and are passed directly to the ACSE user.

9.2.2 Protocol

As we have observed above, CASE provides commonly used Application Layer functions. The intention is not only to ensure a common approach to common functions across all of the Application Layer standards but also to put in a 'place holder' which will allow more complex Application Layer structures in the future including changing the Application context on an association.

With these possible future extensions in mind it becomes clear that certain lower layer functions such as connection establishment and release must remain firmly in the control of the CASE ACSE provider.

The CASE ACSE protocol standard, ISO 8650: Part 2, specifies that the implementation of CASE ACSE, called the Association Control Protocol Machine (ACPM) is the sole user of:

(a) P-Connect;

(b) P-Release;

(c) P-U-Abort;

(d) P-P-Abort.

The ACPM does not make use of or in any way constrain the use of the other Presentation Layer services by other ASEs. It has already been observed that the ACSE PCI will be encoded according to its own default Presentation context *if* the use of multiple Presentation contexts has been agreed on the connection, but if only a single context is in use then that single context must include the ACSE encoding rules.

In general, the protocol exchanges are exactly equivalent to the ACSE service exchanges described above and there is an APDU corresponding to each service primitive.

One of the major features of the upper layers, and of CASE ACSE, is the principle of embedding. It was realised during the development of OSI that sequential exchanges at each layer in order to establish connections from Layer 1 up to Layer 7 would result in a very inefficient system. This is because the major time delays are caused by sending PDUs and receiving replies (as opposed to the end-system delays in processing PDUs). Thus a PDU exchange at a layer to establish a connection followed by a PDU exchange at the next higher layer to establish the higher layer connection would cause considerable delays in starting up an Application association. Thus the standards for CASE, Presentation and Session Layer protocols specify that the PDUs for each layer are carried in the user data field of the corresponding lower-layer PDU. This procedure is called 'embedding' and is also followed in the various SASE standards where SASE PDUs are embedded in the user data of corresponding ACSE PDUs. Thus:

(a) A-Associate PDUs are carried in the user data fields of P-Connect primitives (and are thereby embedded in the user data of P-Connect PDUs);

(b) A-Release PDUs are similarly embedded in the P-Release primitives;

(c) A-Abort PDUs are embedded in P-U-Abort primitives;

(d) A-P-Abort PDUs are also embedded in P-U-Abort primitives.

Because embedding takes place, A-Release APDUs can only be issued when the Session Layer requirements for S-Release are met, i.e. the sender possesses the necessary tokens.

Collisions between ACSE APDUs are resolved implicitly by the collision resolution procedures of the Session Layer since the Session Layer will discard SPDUs and with them the corresponding Presentation and Application PDUs.

The parameters for the ACSE APDUs are in general exactly those from the corresponding ACSE service primitives. The A-Associate APDU additionally contains a 'version number' parameter which for the protocol defined in IS 8650: Part 2 is set to the value '1'. Enhancements and Addenda to the basic protocol will define further values for this parameter.

The only other parameter which does not correspond exactly to the equivalent service parameter is the Presentation context. The initiating ACPM adds its own context to the list of Presentation contexts supplied by the ACSE user if multiple Presentation contexts are to be used and the responding ACPM will remove the ACSE Presentation context from the list before it is passed to the responding ACSE user. The ACPM will also set the 'Result' and 'Source' parameters of Presentation release and abort primitives appropriately.

There is a conformance statement for ACSE, as with all other OSI protocols. The static conformance requirements are simply that the implementation must be able to act as initiator or responder of associations (or both) and support the defined encodings for PCI. The Protocol Implementation Conformance Statement (PICS) must state whether the implementation can act as initiator or responder of associations (or both).

9.3   COMMITMENT, CONCURRENCY AND RECOVERY (CCR)

This section describes the features of the Commitment, Concurrency and Recovery (CCR) Common Application Service Element (CASE). This is the technical term for the standardisation provided by ISO 8649: Part 3 and ISO 8650: Part 3, two important Layer 7 OSI Standards.

9.3.1. History and objectives

The use of 'two-phase commitment' to handle 'simultaneous' updates of two or more separate resources has been known for many years in database work, and has become a common feature of 'transaction' protocols.

Much of the early impetus in ISO came from the needs of the JTM work (see Chapter 11), where the distributed nature of JTM activity led to a requirement for commitment. The JTM work also emphasised the need for well-defined protection against crashes of systems, and the interaction of this with 'commitment' handshakes.

Finally, the crucial interaction with concurrency controls was recognised and the term 'CCR' – Commitment, Concurrency, and Recovery – control was born. The CCR service and protocol are designed to ensure successful co-ordination and completion of activities distributed across several open systems, taking into account the possibility of network failures and system crashes.

CCR is often described as co-ordinating across several connections synchronisation of communications activity which, for a single connection, is handled by the Session Layer. A dual, but perhaps more useful, focus, is to regard it as co-ordinating the information processing at the nodes linked by the connections.

The work neared stability at the end of 1983, but at this time there was a growing interest in use of CCR for other applications – transaction processing, remote database access, directory updates, which introduced elements of instability.

At the time of writing this chapter (June 1986) the final form of the initial CCR Standard is not certain. There is also instability in the precise way the CCR specifications are to be included in other standardisation work, revolving around the so-called 'co-operating main service' issue.

This section describes the main features of CCR as it existed in the second Draft International Standard.

9.3.2 Application of CCR

CCR is primarily concerned with exception cases. It is concerned with what happens if a crash (loss of information, release of concurrency controls) occurs at critical points in an activity, or if one update succeeds but a related update cannot be done because the network has gone down.

This means that almost any application using CCR can be run without CCR, *and will work much of the time*. Nonetheless, for almost all applications, use of CCR gives a degree of reliability which is generally highly desirable.

The following are the main areas where use of CCR, or some CCR-like handshake, is highly desirable:

(a) no-loss, no-duplication transfer of material from system A to system B, without requiring an indefinite retention of knowledge of completed transfers or the intervention of human intelligence on crashes;

• (b) up-dates to two different systems by a third party, where it is essential for neither or both updates to occur before other users access the affected information;

(c) any remote operation where guarantees are required of precisely one performance of the operation, without requiring an indefinite retention of knowledge of completed operations or the intervention of human intelligence on crashes.

The CCR Standards are used (mandatory) for JTM (see Chapter 11), and (optionally) for FTAM (see Chapter 10), and have been proposed for use in Standards being developed for remote database access, reliable transfer service, transaction processing, and common directory services.

9.3.3 What is Commitment, Concurrency, and Recovery?

In a simple protocol, an initiator requests an action and a responder either performs the action and acknowledges it or refuses to do it while providing an appropriate diagnostic. This simple protocol suffers from two serious flaws for general use. These flaws arise firstly from crashes of one or another system and, secondly, from the need for an initiator to work with several responders simultaneously, e.g. to debit one bank account and credit another.

The first flaw, then, is concerned with crashes. Suppose the initiator receives no reply in the simple protocol outlined above. In connectionless operation, this usually means no reply in some finite time; in connection-orientated operation, this usually means some form of reset or disconnect or provider-generated abort. If the crash was due to a network failure, certain classes of connection-orientated Transport service permit recovery. If, however, one or another end-system or end-application crashed (described in the CCR Standard as an Application failure), the Transport Layer offers no help. In this case, the initiator does not know whether the action was performed but the crash lost the acknowledgement or whether the action was refused but the diagnostic was lost, or whether the action request was lost. (Figure 9.1 shows these options.)

A simple 'recovery' protocol puts an identifier on the request. Following the failure case, the initiator retries the action with the same identifier, and the responder *who has remembered the identifier* (on disk) detects a duplicate request and reacts accordingly.

However, for how long must the identifier be remembered? One protocol says 'until a different identifier comes from the same source'. This works

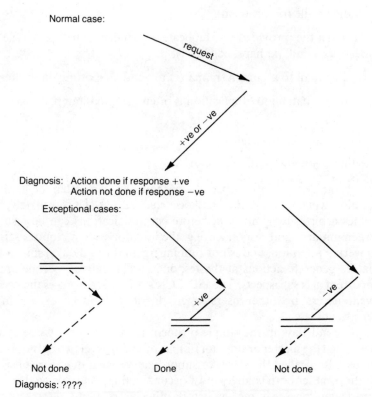

Normal case:

request

+ve or −ve

Diagnosis: Action done if response +ve
Action not done if response −ve

Exceptional cases:

+ve

−ve

Not done   Done   Not done

Diagnosis: ????

Fig. 9.1 Problem in a simple protocol.

fairly well unless you want multiple simultaneous activities between the two, but still involves holding the information indefinitely following the 'last' operation.

'Indefinitely' needs qualifying. What we really require is 'for a period which is long compared with the expected recovery time of the initiator'. Now suppose we guess wrong, and it is too short. The 'reliability' breaks down and we are back to humans sorting the mess out. Depending on the Application, the risk of loss of reliability may need to be low or very low, but in almost all cases it is important for the breakdown to be reliably *detected* and signalled to a human. Some of these problems will be clearer when heuristic commitment is discussed.

Many experts believe that the occasional duplication of an action is inherent in protocol design. *This is not so.* The ISO CCR Standard, if incorporated in an Application, provides 'no-loss, no-duplication' operation. The following actions are ones where risk of duplication ranges from 'no problem' to 'intolerable':

(a) reading a file (no problem);

(b) writing a file (provided the duplicate action occurs before the file has been seen and, perhaps, edited ...);

(c) sending mail to a human (humans are good at spotting duplicates);

(d) running a batch job (duplication is often accepted, but has nasty side-effects);

(e) appending to a file (ugh!);

(f) debiting a bank account (nasty).

In general, Application-specific mechanisms can often be used to determine whether an action has occurred or not. Unfortunately, this becomes arbitrarily complex if, before communications can be resumed between initiator and responder (e.g. the initiator system requires serious engineering attention, and is not running again for two days) some other initiators generate actions at the responder. This relates to concurrency (concurrent access) aspects. The ISO CCR standard recognises the need to prevent access by other users when distributed resources are in an inconsistent state.

The second flaw in the simple protocol is exemplified by one system (often called the master or superior) initiating changes on two or more other systems (often called the slaves or subordinates), such that, for consistent operation, either both changes must occur simultaneously or neither must occur. (Simultaneously means 'before other users access the resource' – concurrency again.) A typical example is debiting an account at system A and crediting that money to an account at system B. In the simple protocol, the master 'commits' itself (to the possibility that the action will happen) when it makes the request (see Fig. 9.2).

The ISO CCR protocol ensures that the master retains control. In what is called 'Phase I' of the CCR handshake, the superior initiates the action and receives a commitment by both subordinates to perform it (or refusal by one or more of them). At this stage, concurrency controls are in place to prevent the changes being apparent to other users. The subordinate undertakes to monitor actions by other users to ensure that the changes to which it is committed will continue to be possible, and that 'rollback' to the initial state *also* remains possible (because the master is not yet committed to the action). It is important here to note that these requirements do *not* in general mean that other users are completely locked out of the resource. Considerable flexibility can be adopted over permitting the use of uncommitted data – particularly if the second use is subject to CCR and the custodian of the data (the subordinate) can still ensure rollback of the

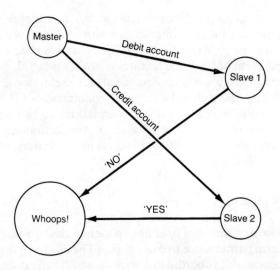

Fig. 9.2 Updating two slaves. Now assume a network crash before Slave 2 can be debited back to its original state. And another user withdraws the cash! The reverse situation can be almost as bad for large sums – thousands of pounds of interest are lost!

second use if the expected commitment of the first change is not forthcoming.

Again, a good example comes from debiting and crediting a bank account, where the unaffected balance can be available for other users at all times (see section 9.3.7). Another example comes from commitment to a credit card debit, when a hotel verifies your credit card when you check in. For example, on checking into a hotel, the hotel (the commitment master) asks the credit card company: 'Will you commit to making a debit for up to this amount?' The credit card company says: 'Yes', and provides an 'atomic action identifier' (authorisation code). At the end of your stay, the hotel commits the debit by sending-in the bank copy with the 'atomic action identifier' on it. Until this action is committed (to the same or a lesser amount), the credit card company avoids any additional credit commitments that would place you over your credit limit.

If one or more subordinates refuse an action, the master (using ISO CCR) will (in what is called 'phase II') order rollback, restoring all resources to their initial state, *and releasing concurrency controls*. If, on the other hand, all subordinates have offered commitment, the master will, in phase II, 'order commitment', that is to say, require the action to occur *and concurrency controls to be released*.

The extra handshakes (protocol) needed to mend each of the two flaws in the simple protocol, and their interactions with concurrency controls, are the same for each flaw. Thus the ISO CCR control protocol is a single Standard. Separation of the three features is not meaningful.

The CCR mechanisms (or their equivalent) are useful (some would say essential) whenever more than two parties communicate. These mechanisms are also useful (some would say essential) in the two-party case if Application-independent recovery is required. Application-*dependent* recovery is often extremely difficult (due to access by other users) – even when using human intelligence.

9.3.4 The CCR exchanges

The CCR service and protocol operates on each of the two-way communications that form part of some tree of activity. The master at the top of the tree communicates with subordinates, each of which can act as a superior to further subordinates, and so on. The CCR standard defines the protocol (and corresponding local actions) between a superior and a subordinate. The local actions for a superior specify rules that link the exchanges with all its subordinates (and its own superior, if any), to ensure that the master retains control at all times. The CCR atomic action tree is shown in Fig. 9.3.

The term *atomic action* describes the piece of work performed during one invocation of CCR. The action is called atomic because, *to an outside observer*, either all the changes (on all systems) required by the action occur,

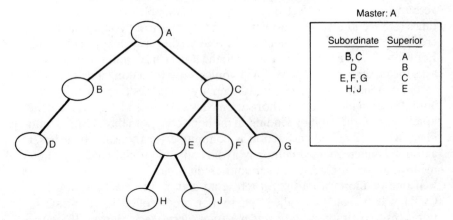

Master: A

Subordinate	Superior
B, C	A
D	B
E, F, G	C
H, J	E

Fig. 9.3 A CCR tree. The distributed application in this example involves the application-entity instances A, B, C, D, E, F, G, H and J. These may all be different application-entities on different open systems, or may have some degree of overlap.

or none of them occur (the action is rolled-back by the master). Ensuring that an action is atomic is dependent on the co-ordinated application and release of concurrency controls.

On any one CCR communication between a superior and a single subordinate, the following events occur:

(a) the superior issues a C-BEGIN primitive (with an atomic action identifier) in order to mark the start of an atomic action.

(b) Application-specific exchanges occur, defining the action. During this phase, any data used for the action has concurrency controls, which are associated with the atomic action identifier, applied.

(c) the superior explicitly (with a C-PREPARE primitive) or implicitly (by an Application-specific exchange) indicates that the intended action is complete and that it wishes to know whether the subordinate is prepared to commit.

(d) The subordinate offers commitment with a C-READY primitive (or refuses it with a C-REFUSE and a diagnostic, either now or at any earlier time). From this point on, the subordinate is required to ensure that both commitment to the action *and* rollback continue to be possible. This is achieved by applying concurrency controls to all affected resources.

(e) The superior orders commitment by a C-COMMIT primitive, or rollback (restoration of the initial state of resources) by a C-ROLLBACK primitive.

(f) The C-COMMIT or C-ROLLBACK is a confirmed service, the response/confirm providing an acknowledgement by the subordinate to the superior that all concurrency controls have been released, and the subordinate has 'forgotten' the action. When this is received, the superior can also 'forget' the action and, if it wishes, can even re-use the atomic action identifier in a subsequent action. Note the critical nature of the final message in telling the master his responsibilities are at an end.

Figure 9.4 shows this sequence.

9.3.5 Recovery procedures

Once C-BEGIN has been issued, a conforming superior is *required* to complete the atomic action with a rollback or commit. This is to ensure release of resources by the subordinate. Thus, following a crash (of either end) the superior tries (and keeps trying) to restart the atomic action, quoting the atomic action ID, in a C-RESTART primitive.

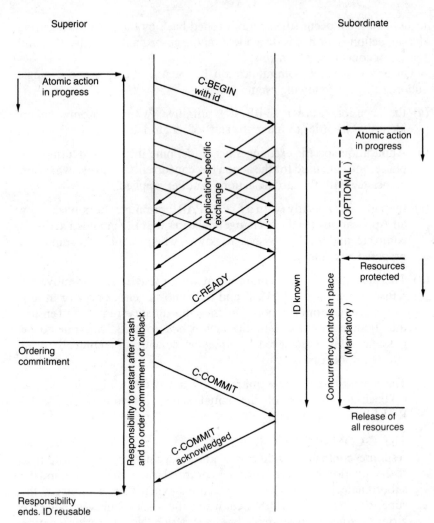

Fig. 9.4 Basic CCR handshake.

The ID would be unknown if the crash occurred prior to receipt of the C-BEGIN or after issue of the C-COMMIT (or C-ROLLBACK) acknowledgement. The superior can distinguish these two cases because it will have recorded (on disk) its decision to commit. An update at this point is essential to the correctness of the protocol.

At the subordinate end, resources will normally be protected (and the protection recorded on disk to guard against crashes) as they are used in the action. The *minimum* requirement, however, is to record the protection of resources when C-READY is issued.

The CCR specification ensures that its requirements are the *minimum* necessary number of disk updates to ensure fully reliable operation of CCR, no matter when crashes occur.

9.3.6 Heuristic commitment or rollback

With this simple protocol, the actions that a subordinate implementation can take following a 'crash' are very constrained. Prior to offering commitment, rollback (release of resources in the initial state) is always possible, either immediately or if the superior takes too long to restart the action. This subordinate is simply 'pretending' that the C-Begin was not received, or that a C-Refuse was issued, but lost in the crash.

However, once commitment has been offered, the affected resources must be protected against other users until the superior restarts. Thus, in the simple protocol, a conforming implementation is expected to retain the associated concurrency controls for an effectively indefinite period.

What is missing from this simple protocol is the important concept of heuristic commitment.

Heuristic commitment (or rollback) permits a subordinate who has offered commitment, and who subsequently loses communication with its superior or crashes, to decide unilaterally either to commit – heuristic commitment – (or to rollback – heuristic rollback) if the superior fails to restart soon enough.

This is called heuristic commitment or rollback because the subordinate must *guess* whether:

(a) the superior is more likely to order commitment than rollback;

(b) whether guessing wrong does more damage in the commitment or the rollback case.

The decision is normally expected to be taken by a human being!

The disadvantage of heuristic decision is that the whole CCR service *fails* if the guess is wrong – the atomic nature of the action is lost, and distributed resources can be left in an inconsistent state (e.g. a bank account has been debited, but no account has been credited), and loss or duplication of actions can arise.

The need for a heuristic decision arises from the undesirability of applying concurrency controls to a resource for an indefinite period (especially if 'locking' is used – see section 9.3.7). In general, the need for heuristic action will depend on the demands of other users to access the resource and the importance of meeting those demands.

In the full OSI CCR protocol, the superior, on C-BEGIN can:

(a) say how long it requires the subordinate to wait (after a C-READY) before indulging in heuristic commitment or rollback; the subordinate accepts this constraint by issuing C-READY – if the constraint is unacceptable, it issues C-REFUSE;

(b) say 'please make the heuristic action (if taken) commitment', or 'please make the heuristic action (if taken) rollback', or 'you can choose'.

It is important to distinguish release of resources (concurrency control) on heuristic commitment or rollback from the complete amnesia which occurs on a normal commitment or rollback. The subordinate who takes a heuristic decision has a responsibility to retain knowledge of the atomic action, and the decision it took, until the master has attempted restart, been told a heuristic occurred, *and told the subordinate it has received that information*. Only then can the subordinate forget the identifier, and only when the master has been told this has been done do the master's responsibilities end. Thus, heuristic commitment adds significant complication and extra handshakes to the restart procedures.

9.3.7 Concurrency controls

A very crude approach to protecting resources to permit later orders to commit or rollback is to apply 'locks' which prevent access by all other users. ISO CCR recognises that this is too simplistic an approach. Consider, for example, debiting an account. Provided there is still money left, other debits should clearly be allowed 'simultaneously' (while commitment is offered), and certainly credits should be allowed!

ISO CCR defines 'the period of use of a datum' by an atomic action, which is the time from first to last reference to the datum. The CCR requirements are simple:

(a) an implementation is required to rollback an atomic action if, during the period of use of a datum by the atomic action, the datum is changed by other actions; and

(b) an implementation is required to prevent an atomic action from committing if it has used any datum that has been changed by an atomic action that has not yet committed.

These requirements can be satisfied in many ways. For example, if a datum is locked (preventing change by other atomic actions) from first use to the end of phase I, (a) above is automatically satisfied. It can be satisfied equally by issuing a C-REFUSE during an action if (a) is found to have been violated.

In the case of (b), locking (preventing another action from accessing the resource once commitment has been offered) is one possibility, but another would be to allow other actions to proceed on an assumption that commitment will occur, and to delay the offer of commitment for these actions until the earlier commitment arrives. Yet a third variant covers the case of debiting an account. Here the data (the balance) can be regarded as two separate pieces of data; one is the residue after debit, which is unaffected by the atomic action, and the other is the bit being (potentially) removed, which disappears on commitment or is reinstated on rollback. Any subsequent debits that can be supported by operations on the stable residue part (the account has enough money in it for both debits) can now proceed to commitment.

9.3.8 Additional features

Three aspects of the CCR operation are worth discussing but are not covered by the intitial CCR Standard.

The first aspect is (global) checkpointing, which covers the totality of the actions up to some moment within the entire atomic action tree. (This should not be confused with checkpointing of a one-way flow of data between two parties; it is more both powerful and more 'expensive' in round-trip exchanges.) This is important for long atomic actions in order to prevent crashes from having to be recovered by starting the entire atomic action again from the beginning.

The second aspect is that of nested atomic actions. Consider, for example, a series of atomic actions ordering a list of goods from a series of warehouses, and perhaps involving negotiating with a transport contractor (for each one). The ordering of each item in the list involves a commitment handshake to handle the correlation of the transport contractor commitment and the warehouse commitment, and hence can usefully be structured as an atomic action. However, the customer may wish to abandon the entire list unless he gets about 80% of it satisfied from this supplier. Thus, he wishes to have an outer atomic action wrapped around the series of inner ones.

The final aspect involves reopening phase I of an atomic action. The point here is best illustrated by the hotel and credit card example. In phase I, the credit card company commits to a certain size of debit. The actual debit is, however, less. In the real world this is handled by indicating the lesser debit when ordering commitment in phase II.

With OSI CCR, however, the choice is to either:

(a) commit the greater debit, then do another atomic action to add a credit for the difference; or

(b) rollback the first atomic action, then do the actual debit as a new atomic action, risking other users getting in-between, and hence failure of the second debit.

What is required is the opportunity to renegotiate a commitment offer (to the lesser debit) while retaining the possibility of committing the earlier offer.

9.3.9 CCR service primitives and parameters

The CCR service primitives are listed below:

Primitive	Issued by	Confirmed?
C-BEGIN	Superior	No
C-PREPARE (optional)	Superior	No
C-READY	Subordinate	No
C-REFUSE	Subordinate	No
C-COMMIT	Superior	Yes
C-ROLLBACK	Superior	Yes
C-RESTART	Superior or Subordinate (only if communications have not failed)	Yes

The parameters of C-BEGIN are:

Atomic action identifier – an Application-entity-title that unambiguously identifies the master, and a suffix that is a character string unambiguously identifying the action among all current actions with the same master.

Branch identifier – the Application-entity-title that unambiguously identifies the superior (and which is carried in the Association Control CASE protocol) together with a suffix that is a character string unambiguously identifying this branch of the atomic action tree among all branches of the same atomic action with the same superior.

Atomic action timer – (optional and advisory) – a signed integer value (N) that warns the subordinate that the superior intends to rollback the atomic action if it is not completed in $2^{**}N$ seconds.

Heuristic timer – (optional) if present, specifies the time the subordinate is required to wait after offering commitment before it is permitted to perform heuristic commitment or rollback. It is either 'indefinite' or a value in seconds obtained like the atomic action timer. (Note that the timer is *permissive*. Heuristic commitment is not *required* to occur after its expiry.)

Heuristic-decision – (optional) if present, constrains any heuristic decision to be the one it specifies (i.e. either Commit or Rollback).

User data – carries information determined by the SASE with which CCR is incorporated.

The C-PREPARE, C-READY, and C-REFUSE primitives carry only user data. The user data on C-REFUSE is expected to carry structured diagnostic information.

The C-COMMIT and C-ROLLBACK primitives carry no parameters.

The C-RESTART request and indication carries the following parameters:

Atomic action identifier – the same as C-BEGIN.

Branch identifier – the same as C-BEGIN.

Restart timer – (optional) the same form as the atomic action timer; the superior intends to break the association (and attempt the restart later) if a reply is not received within the stated time.

Resumption point – ACTION if the superior has not issued (prior to the crash) a C-COMMIT or C-ROLLBACK, otherwise it is COMMIT or ROLLBACK respectively.

User data

The C-RESTART response and confirm carries only:

Resumption point – DONE, RETRYLATER, REFUSED, COMMITTED, ROLLEDBACK, MIXED or ACTION.

A response of DONE indicates either that the C-BEGIN was not received or a C-ROLLBACK or C-COMMIT response has been issued (atomic action ID unknown), or that the C-RESTART ordered COMMIT or ROLLBACK and commitment or rollback (respectively) has occurred (either now or by an earlier heuristic).

The response RETRYLATER indicates that restart is not possible at this time.

A response of REFUSED indicates that the subordinate has previously issued a C-REFUSE for this atomic action.

A response of COMMITTED (or ROLLEDBACK) means that ROLLBACK (or COMMIT) has been ordered, but that a previous heuristic commitment (or rollback) decision had been taken (respectively).

A response of MIXED means that some subordinates of the subordinate have heuristically committed and others have heuristically rolled-back, or that a partial heuristic has been taken, with some resources still available for commitment or rollback.

Following a COMMITTED (or ROLLEDBACK) response, the superior issues a C-COMMIT (or C-ROLLBACK), respectively. (This final handshake is needed to ensure that the superior has received notification of the incorrect heuristic decision before the subordinate forgets the atomic action.)

Following a MIXED response, the superior issues either a C-COMMIT or C-ROLLBACK to commit or rollback any remaining resources, but with a 'MIXED' parameter present to acknowledge receipt of the information that heuristics have occurred.

Following a response of ACTION, the atomic action is restarted, either from the beginning or from an Application-specific checkpoint, or from a CCR global checkpoint.

9.3.10 The CCR protocol

The CCR specifications contain the following features:

(a) an abstract syntax defined using ASN.1 (see Chapter 8) for a set of messages capable of carrying the information on the CCR service primitives;

(b) a specification of the semantics and local actions associated with these messages, and the order in which they can be sent;

(c) a means of transferring these messages using the Presentation service.

The correspondence between the messages for each service primitive and the Presentation service primitive used to carry them is shown below:

CCR service	*Presentation service*
C-BEGIN	P-SYNC-MAJOR
C-PREPARE	P-TYPED-DATA
C-READY	P-TYPED-DATA
C-REFUSE	P-RESYNCHRONISE (ABANDON)
C-ROLLBACK	P-RESYNCHRONISE (ABANDON)
C-COMMIT	P-SYNC-MAJOR
C-RESTART	P-RESYNCHRONISE (RESTART)

9.3.11 Co-operating main services

The inclusion of (c) above in the CCR Standard has been a matter of controversy, because:

(a) the ISO FTAM protocol conveys CCR messages as part of its own PDUs (in order to avoid a round-trip time for FTAM *and* for CCR, and in order to make minimum use of Session features);

(b) extending protocols based on use of S-Activity (as most early CCITT protocols were) by adding the CCR messages as further (optional) fields in the existing protocol provides an easy upgrade path, but adding extra S-SYNC-MAJORS is more or less impossible;

(c) one view of upper layer architecture is that CASE Standards have no business specifying the means of transfer; it is for the designer of the SASE making use of the CASE to put together the total protocol in whatever way he thinks is best, using all available tools.

In partial recognition of these points, an annex to the CCR Standard describes the use of CCR by 'a cooperating main service', which carries CCR messages in fields of its own PDUs, as FTAM does.

9.3.12 Analysing 'reliable' protocols

This part of the text introduces the reader to the technique of analysing protocols which claim to be 'reliable'. It is not essential to an understanding of CCR, and can be omitted on a first reading.

The reader will encounter, and may become involved in designing, 'reliable' protocols.

It is important to be capable of analysing such protocols to see if the stated goals are being achieved.

Analysis is relatively simple provided a suitable methodology is followed. The following steps are needed:

(a) identify clearly the incoming protocol events which produce changes on disk (memory which survives crashes), and any resulting outgoing events;

(b) draw a time-sequence diagram similar to that in Fig. 9.5 marking these points clearly. (The six horizontal arrows are the points for the OSI CCR protocol.)

(c) consider crashes (cessation of activity and reversion to the last horizontal arrow) at all possible timelines, and see what the protocol does.

Figure 9.5 shows the time-lines (dashed lines) for OSI CCR, marked 'A' to 'E'.

It is important to note that for CCR all actions depend only on the main CCR protocol. Other reliable protocols will place semantics on P-Disconnect or P-Release messages. If so, these must be included in the time sequence diagram.

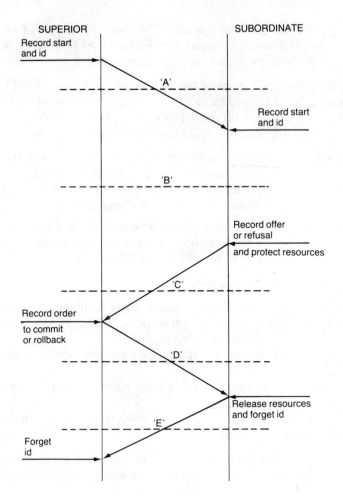

Fig. 9.5 CCR analysis.

A crucial point in any analysis is examination of the *last* message. This message always has the property that if the crash occurs after its issue and before its receipt (line 'E') this is undetected by the sender of it. Moreover, as he has no remaining memory (if he has, your time sequence diagram is not complete), any information on it is *lost* and cannot be repeated following a restart. This leads to an interesting 'principle' or 'law' for reliable protocol endings – 'A correct reliable protocol cannot carry any information on its last message'. This often provides a very quick rule of thumb for saying 'protocol X does not work', even without detailed analysis.

Let us now examine CCR. (Note, we are only outlining the 'proof'. A more rigorous treatment would also examine the restart exchanges and the heuristic commitment exchanges.) Referring to Fig. 9.5:

At point 'A':   The superior issues C-RESTART (ACTION); the subordinate says 'DONE' (the ID is unknown); the superior starts the action again with C-BEGIN. OK.

At point 'B':   The superior issues C-RESTART (ACTION); the subordinate replies C-RESTART (ACTION), and the action begins again (from the start or from an Application-specification checkpoint). OK.

At point 'C':   The superior issues C-RESTART (ACTION); the subordinate repeats any refusal or restarts the action, typically from a checkpoint near or at the end. OK.

At point 'D':   The superior issues C-RESTART (COMMIT) or (ROLLBACK); the subordinate performs commitment or rollback as if it were a C-COMMIT or C-ROLLBACK. OK.

At point 'E':   The superior issues C-RESTART (COMMIT) or (ROLLBACK); the subordinate says 'DONE' (the ID is unknown); the superior has recorded his order to commit or rollback, so can distinguish this from case 'A', and tidies up. OK.

A further complication in CCR is to consider the records which need to be kept by intermediate nodes in the CCR tree. The absence of 'OFFERED' on C-RESTART responses is to avoid forcing further disk updates in 'normal' (non-crash) cases. (Inefficiency in the restart case is several orders of magnitude less important.)

The above analysis was for a protocol which works. Now let us try one that has some slight flaws!

We use the 'activity' functional unit of session, and what we are trying to do is to reliably transfer electronic mail through mail relays to a destination.

We start with P-ACTIVITY-BEGIN, ship the mail, end with a P-ACTIVITY-END request and get back either:

P-U-EXCEPTION-REPORT
          (problems)

or   P-ACTIVITY-END response/confirm
          (OK)

or   timer expiry (no response)

If 'OK', we issue P-DISCONNECT, otherwise we issue P-ACTIVITY-DISCARD and try another route.

On crashes between P-ACTIVITY-BEGIN and the P-ACTIVITY-END response/confirm, we try again with P-ACTIVITY-RESUME, but may also send on another route if we can't get through for a long time, intending to do P-ACTIVITY-DISCARD following the eventual P-ACTIVITY-RESUME.

The mail is forwarded by the subordinate when a P-ACTIVITY-END response is given, and the ID is forgotten on a received P-DISCONNECT.

This is the basis of the CCITT X.410 protocol, although it has been simplified slightly in this description.

Figure 9.6 shows the recording and time-sequence. Now for the analysis of crashes:

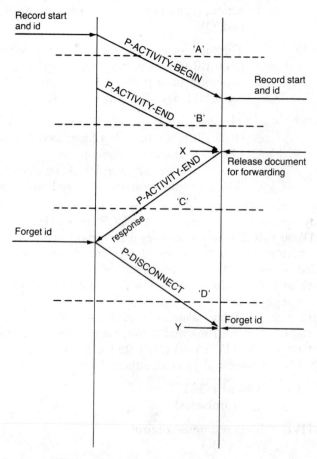

Fig. 9.6 Another reliable protocol.

At point 'A':	The superior sends a P-ACTIVITY-RESUME, it is treated as a P-ACTIVITY-BEGIN because the ID is unknown. OK.
At point 'B':	Again, P-ACTIVITY-RESUME, picks up all right. OK.
At point 'C':	Again, P-ACTIVITY-RESUME, mail has been forwarded, but ID is still known, so S-U-EXCEPTION-REPORT (or continue and discard the duplicate). OK.
At point 'D':	ID is left dangling – has to be remembered indefinitely, or deleted by other means. (Note that the subordinate can't distinguish 'D' from 'C'.) Bad (but not, of course, disastrous).

Now consider again the crash *at 'C'* (no response for a long time, say), and the issue of a P-ACTIVITY-DISCARD by the superior (now, perhaps colliding with the late P-ACTIVITY-END response, or later), with the mail now being shipped on an alternative route. We now have *duplication* of the mail.

Comparing this with the CCR handshake we can see the cause of the problems. The first is the release of concurrency controls (sending the mail onward) at point 'X' instead of point 'Y' (the master lost control, and so potential duplication was introduced), and the second is the absence of confirmation at point 'Y', leading to the occasional 'dangling ID'.

One final remark is relevant. We have seen here the *occasional* dangling ID, and the *occasional* duplication. For *some* purposes, even occasional duplication in a supposedly 'reliable' protocol would be wholly unacceptable. For human mail, it is clearly all right. Thus formal analysis must be tempered by 'fitness for purpose' considerations.

The above examples should now enable the reader to examine any 'reliable' protocol, to find if it has any flaws. As a quick check:

(a) Does it have the double handshake (keeping control with the master) before release of the material?

(b) Does it have the one with restart responsibility forgetting the ID *last*?

(c) Does it have information on the last message which affects the reliable protocol?

If you get 'Yes', 'yes', 'no', there is a good chance it works! Otherwise, start looking for the flaws!

9.3.13 Implementation

The CCR protocol is very simple to implement from the communications point of view. All the problems are concerned with:

(a) getting reliable storage of information which must be retained across crashes; and

(b) managing the protected resources which are to be committed or rolled-back.

The first problem stems from the fact that the master *must* have a disk record of its responsibilities to clear the action before the subordinate ties-up resources by making *its* disk record. On most 1980s operating systems, this involves writing to a disk file, closing the file, *and ensuring operating systems buffers are written up* before sending C-BEGIN. *Updating* this stored information by an intermediate node in the CCR atomic action tree requires updates of records for different connections to be made as an atomic action – all or none. This often requires considerable care, and use of cyclic version numbers for file-names holding the records. There are no insuperable difficulties, but understanding and care is needed if the crucial reliability features of CCR are not to be lost by an incorrect implemention.

The second problem is much more difficult to handle unless support for CCR is built deeply into the operating system. It requires in the simplest case mechanisms to reinstate 'locks' following recovery from crashes; in the case of flexible application of concurrency controls (see section 9.3.7 and the bank-balance discussion) it can require high-resource and application-dependent gate-keeping of access which can become extremely complex, particularly if use of uncommitted resources is to be permitted.

The implementation of CCR is considerably eased, however, if it is being considered in relation to a specific application which is the only accessor of the relevant resources. In this case, support of the CCR controls can be built into the application, and some of the problems evaporate.

9.3.1 Conclusion

The OSI CCR Standards provide a very important tool for application designers.

As more use is made of OSI CCR, and experience gained with it, it can be expected that there will be significant additions to what is, in the initial Standards, an essentially simple Standard.

PART 4

COMMUNICATION BETWEEN OPEN SYSTEMS: SPECIFIC SERVICES

CHAPTER 10

File Transfer, Access and Management

This chapter describes the features of the file transfer access and management (FTAM) specific application service element (SASE). This is the technical term for the standardisation provided by ISO 8571: Parts 1 to 4, an important Layer-7 OSI Standard.

10.1 HISTORY AND OBJECTIVES

When general-purpose computer networks were developed, one of the first facilities to be supported by vendor-specific protocols was the transfer of disk files from one computer to another. In many cases the protocol was concerned solely with moving complete files, but usually included the ability for an initiator to:

(a) read (and perhaps delete after reading) a remote file;

(b) write to an existing remote file (error if no file exists);

(c) write to a new remote file (error if the file already exists);

(d) write to a remote file, overwriting any existing one or creating a new one;

(e) perhaps append, with similar options to 'write' above.

These protocols normally sat 'monolithically' on top of a basic communications service. In particular, they defined their own encodings for both their own protocol and for a very limited range of document types (typically ASCII text and binary). They had no OSI Transport Layer beneath them to protect and recover from network crashes, so they almost always included checkpointing at the file transfer level to provide recovery by retransmission from the last checkpoint. (Implementations frequently held checkpointing data in main memory, thus protecting against network failure, but providing no protection from host failures.)

The ISO work not only attempted to take the best features of existing *file transfer* protocols, but also broadened its aims. It recognised:

(a) the need to support *file access* (reading or writing random parts of a suitably structured file);

(b) the need to support a wide range of document structures and file types, not just text and binary, with access to parts of complex structures;

(c) the need to support reading and writing (where applicable) of 'attributes' of a file (e.g. such things as 'date last written', 'access control' and so on). This, together with creation and deletion of files, is called *file management*;

(d) the need to encourage/require implementors to use secondary storage for checkpointing in order to protect against host crashes.

The resulting file *transfer*, *access*, and *management* (FTAM) Standards are quite large and complex, with a lot of functionality, some of which is difficult to implement with operating systems of the 1980s. A minimum required implementation is, however, specified, and this is implementable on 1980s operating systems. Some of the problems of full implementation are discussed in this text.

10.2 PHILOSOPHY OF FTAM STANDARDISATION

The nature of filing systems, and in particular the forms of access control and other information held with a file, is very variable across different computer vendors, and sometimes even within a single vendor.

Similarly, the precise nature of files which can be held on a system, and the facilities for segmented, indexed, and random access files are very diverse. Even for the so-called 'simple text' or 'simple binary' file, computer vendors differ. For example:

(a) a binary 'unit' of 8 bits, 16 bits, or 6 bits;

(b) unlimited or limited line lengths in text files;

(c) character-set differences in text files.

Another important difference between systems is in the directory structure supported, ranging from simple one-level systems to elaborate tree structures with cross-linkages and aliasing (i.e. giving alternative names to a file).

In order to produce a standardised protocol for FTAM it is necessary to form a 'model' of a filestore, its directory structure, and the files within it. This is called, by FTAM, *the virtual filestore*, and the FTAM protocol is defined entirely in terms of operations on this virtual filestore.

An FTAM implementation needs first to determine a sensible mapping between the virtual filestore and the real filestore (for file attributes, for file types, and for directory structure). The defined actions of the protocol on the virtual filestore can then be interpreted as actions on the real filestore, and implemented as such (see Fig. 10.1).

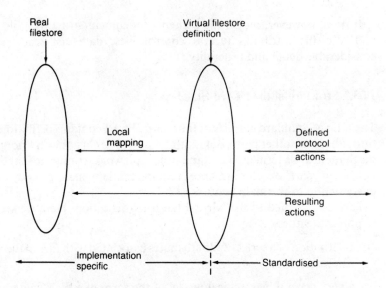

Fig. 10.1 The real and virtual filestore.

Now, suppose the virtual filestore definition is kept very simple. It will be (relatively) easy for implementors to map all the features of the virtual filestore on to their real filestore, and to fully and completely support the FTAM Standard. Unfortunately, many real facilities and real local file types would then find no equivalent in the FTAM virtual filestore and, hence, could not be supported by the FTAM protocol. This would inevitably lead to the continued use of vendor-specific protocols for operation between two like machines, with the FTAM protocol regarded as considerably inferior.

Suppose, on the other hand, the FTAM virtual filestore contained a wide range of functionality (greater than any actual real system), all supported by protocol. In this case, it would be (relatively) easy to use the FTAM protocol as effectively as a vendor-specific protocol, but hard, perhaps impossible, to produce a full implementation of the Standards.

The philosophy underlying FTAM standardisation leans much more in the latter direction (richness of facilities) than in the former (easy implementation).

10.3 OVERVIEW OF FTAM

The initial Standards have *no* detailed support for a filestore directory structure (the text of this Standard indicates likely future extensions to support directory structures). The implicit structure is a monolithic single-level naming of files.

It does, however, model file contents (document structures, file types) and also file attributes (access control lists, date last read, etc.) in considerable detail and generality.

10.3.1 Structure of the FTAM Standard

The FTAM Standard consists of four parts. Part 1 contains definitions used throughout the other parts, but is otherwise a series of annexes ('not part of the Standard' – a term used in standardisation work to indicate that they do not form part of conformance requirements) which give a tutorial introduction to the main features of FTAM.

Part 2 is devoted to the 'virtual filestore' definition. Here we see three important features:

(a) a definition of a series of file attributes associated with files in the virtual filestore;

(b) a very general hierarchical model of the form of a file's contents;

(c) the definition of a number of 'constraint sets' and 'document types' which restrict the generality of (b) above to provide file contents suited (hopefully) to operating systems of the 1980s.

In terms of understanding the features of FTAM, and what can be expected of an FTAM implementation, Part 2 is the most important.

Parts 3 and 4 are the 'service definition' and 'protocol specification' for FTAM. The service definition (Part 3) defines the interactions between the FTAM SASE and the virtual filestore, and between the FTAM SASE and a remote user trying to access the virtual filestore. It specifies the allowed sequences of the interactions, and the parameters of each interaction. For example, it defines an

F-OPEN

service primitive to open a file, which is later followed by an

F-CLOSE

service primitive to close the file.

The parameters of F-OPEN are specified in Part 3 of the Standard.

The protocol specification (Part 4) defines the messages which are transferred across a network to communicate the wishes of an 'initiator' (the remote user SASE) to a 'responder' (the virtual filestore SASE), in response to the issue of service primitives.

10.3.2 OSI service definitions and FTAM

We have seen in Chapter 3 that a service definition is the means whereby a 'higher layer' Standard can invoke or be invoked by procedures specified in some lower-level Standard. The use of this form of specification for FTAM recognises the possibility of later 'value-added' Standards using FTAM as part of the communications infrastructure (see Chapter 7). For most implementers today, however, the importance of Part 3 is the effect of the defined service primitives on the virtual filestore.

The style of OSI service primitive definitions is to define four types of interactions that may occur for each 'service' offered by a protocol. Thus, for the 'open file' service, we have the following events:

(a) a 'Request' service primitive (issued on an implementation-dependent basis) which causes a protocol message to be despatched from the remote system to the system supporting the virtual filestore;

(b) an 'Indication' service primitive, which results from the receipt of the protocol message in (a) by the system supporting the virtual filestore;

(c) a 'Response' service primitive (issued on an implementation-dependent basis, but which is almost the only permitted next action) which causes a protocol message to be sent from the system supporting the virtual filestore to the remote system;

(d) a 'Confirm' service primitive which results from the receipt of the protocol message in (c) by the remote system.

This service primitive notation was covered more fully in Chapter 3.

In general, the parameters of an 'Indication' are identical to those of a 'Request', and the parameters of a 'Confirm' are identical to those of a 'Response'. However, the 'FTAM service provider' (the two FTAM SASEs, together with lower-level entities), is modelled as an *active* entity, so for some 'services', the parameters are *not* the same (particularly where 'negotiation' of support for some feature is involved).

A final point on service primitives is to remember the existence of three types of interaction. A 'Confirmed service' has a Request, an Indication, a Response, and a Confirm, as described above. An 'Unconfirmed service' has only a Request and an Indication – any failure to satisfy the Request is reported, in FTAM, by a general-purpose (service-independent) error reporting mechanism (which itself is an unconfirmed service, initiated by the system supporting the virtual filestore). Finally, a 'provider-generated service' has only an indication *at both ends*; without exception, such services in FTAM are used only to report failures or errors.

10.3.3 The FTAM protocol

Once the service definition of FTAM: Part 3 is understood, the protocol specification of FTAM: Part 4 contains very few surprises. It specifies the detailed form of protocol messages to carry the semantics of the Request and Response primitives, and so is of interest mostly to implementors.

There are two points to be noted concerning Part 4. One is the use of a notation called 'Abstract Syntax Notation One' (ASN.1) (see Chapter 8), and the second is the specification of procedures to provide a 'reliable service', to be contrasted with the main 'user-correctable file service' as it is called by FTAM. The use of ASN.1 is discussed below. The 'reliable service' issues are discussed in section 10.3.8.

10.3.4 Abstract Syntax Notation One

The OSI architecture provides a common lower-level presentation service which determines the precise bit-pattern needed to transfer the semantics of an Application Layer protocol such as FTAM. To use such a service, there needs to be a well-defined notation for a Layer 7 protocol to specify the information content of its PDUs. *Encoding rules* are then applied (as a Layer 6 service) to determine the actual bits used to carry the message. Presentation (Layer 6) Standards permit negotiation of the actual bit-patterns (transfer syntax) to be used, allowing for 'normal' transmission, encryption, compression, or vendor-specific formats; such activity is wholly invisible to FTAM.

The notation used to specify the information content of FTAM messages is ASN.1. It evolved from the Xerox 'Courier' specification, through CCITT Recommendation X.409, and into an ISO Standard. A detailed description of this notation is in Chapter 8. It is, however, broadly equivalent in functionality to the notation available in the Pascal language for type definitions; or the functionality of Bachus-Naur form (BNF). Both these mechanisms (and ASN.1) have certain primitive types (integer, Boolean, character string, etc.) and permit their composition in a variety of ways into data aggregates.

10.3.5 FTAM overview concluded

To understand FTAM we need to:

(a) understand the OSI 'architecture', and in particular the 'service primitives' notation (see earlier chapters);

(b) understand the nature of Layer 6 functionality and the role of ASN.1 (see Chapter 8);

(c) understand the general hierarchical model for a file's contents (see section 10.4);

(d) understand the file attributes specified for the virtual filestore (see section 10.5);

(e) understand the sequence and parameters of the service primitives which provide the interaction between a remote user and the virtual filestore (see section 10.6);

(f) understand the 'constraint sets' and 'document types' defined by FTAM in relation to (c) above (see section 10.7);

(g) understand the nature of the 'reliable service' (see section 10.8).

10.4 FILE CONTENTS

The structure of the contents of a file manipulated by FTAM has to be modelled in some way if operations (e.g. 'locate' (a part of a file) or 'erase' (a part of a file)) are to be given meaning and suitable parameters are to be agreed for operations. We say that the 'access structure' has to be modelled. For many real filing systems, the part of a file used for locate and erase would be termed a 'record', and the file would have a 'flat' structure, consisting of a series of records. 'Locate' would be by position in the series. But is this general enough to cover all cases? Clearly the structures in a database go far beyond this simple model. When is data with a complex structure a database and when is it a file? The answer has little to do with the complexity of the structure, but rather is concerned with the enforcement of integrity rules. And what of spreadsheets? Surely these are simply heavily structured files? And what of 'keyed' record structures, where 'locate' can operate on the value of a specified field of the record? Thus the FTAM standardisation group had to produce a somewhat more general model of a file's contents then simply a flat series of records.

On the other hand there needs to be ways of expressing the constraints on the general model which reduce it to the sort of file structures commonly occurring in real systems.

These problems are resolved using three concepts. First, we have a general model. Then we have a number of *constraint sets* defined which restrict the access structure (but do not constrain the actual data types in the file). Finally, we have a number of *document types* defined which completely define the type of the file's contents by specifying a constraint set and the possible datatypes in the file. The general structure is covered in section 10.4.1. Constraint sets and document types are dealt with in section 10.7.

10.4.1 The general hierarchical model

The FTAM model for a file's contents recognises, for access purposes only, a hierarchical structure (and simplifications of it) for the relationship between pieces of data (data units) in a file. This appears to cover most sorts of access and update mechanisms used for things called files, but clearly falls short of the access and update mechanisms used for things called databases. (The 'remote database access protocol' work was started in 1986, but is not reported on in this text.)

Imagine an arbitrary tree structure with a single root and with a defined order for the nodes hanging below any given node – an ordered rooted tree. Imagine further that the nodes hanging below a node at level n in the tree are *not* required to be at level $n + 1$, but each one can be at level $n + 1, n + 2, n + 3$, etc. (this is the so-called 'long-arc' feature). This is the basic structure with which FTAM operates (see Fig. 10.2).

Every file has to be related to (some subset of) such a structure if the FTAM access primitives are to be applied to the parts of the file.

Given a tree structure of this sort, we generate classes of files by adding the following features:

(a) Each node can have a datatype associated with it called *the node name* ('R', 'A', 'B', 'C', etc, in Fig. 10.2). The form of this datatype is not constrained by the general model; the datatype could vary with the position of the node in the structure, and could, indeed, contain an arbitrarily large amount of information. The node name, if present, can

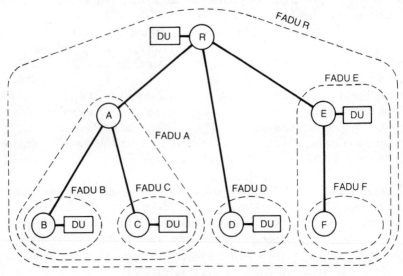

Fig. 10.2 FADU structure.

be used in 'locate' (the only actual restriction on node names is that there should be a defined transfer syntax which specifies a single arbitrary-length bitstring as the encoding of each node name – as opposed to multiple bitstrings; we say the mode name has to be *a single* presentation data value.

(b) Each node may or may not have associated with it another piece of data, called a *data unit*. In general, there is complete freedom on whether a node has an associated data unit or not, although particular constraint sets may, for example, require that all leaf nodes contain units; the data unit is again an arbitrarily complex datatype, but is not restricted to being a *single* presentation data value (the transfer syntax may involve multiple non-self-delimiting strings, and could in theory involve several different encoding rules). This internal structure of a data unit is not visible for access by FTAM primitives.

The final step is to recognise that the FTAM protocol operates almost entirely on *subtrees* (and their associated data). The term *file access data unit* or *FADU* is used to describe the complete information content of a subtree (i.e. its structure, data units, and node names). It is very important not to confuse a 'data unit' and a 'FADU'. In general, a FADU will contain a number of data units. Note that subtrees and, hence, FADUs, are nested. (Figure 10.2 illustrates the FADU structure.)

Subtrees of a tree can, of course, be placed in one-to-one correspondence with nodes by using the root node of a subtree to 'identify' the subtree (or FADU). Thus, one frequently describes the node name of the root node of the subtree corresponding to a FADU as *the FADU name*. A more general term, *FADU identifier* is employed for the parameter used to locate a FADU. A FADU identifier can, in general, take any of the following forms:

(a) A FADU name, if the applicable constraint set requires all node names to be distinct;

(b) a series of FADU names, naming descendants of parents, starting from the root, if the applicable constraint set requires descendants to have unique names;

(c) an integer, identifying a FADU in the so-called 'pre-order traversal sequence'. FTAM defines a linear ordering of nodes (and hence FADUs) in an arbitrary tree (basically, start at the top, and include nodes going down and to the left; go to the right only if you can't go down, and up only if you can't go right; the resulting sequence of nodes is the 'pre-order traversal sequence'). Another option specifies a level, identifying the first node in the preorder traversal sequence at this level;

(d) 'first' – usually this identifies the FADU corresponding to the root of the tree, but a particular document type definition can vary this; for example, for a file constrained to have only a root node with no data unit and Level 1 leaves with data units, the most 'natural' definition of 'first' would be the first Level 1 node, not the root;

(e) 'next' is defined in relation to the currently located FADU and the pre-order traversal sequence;

(f) 'last' and 'previous'; these are analogues to (d) and (e) above respectively, using the pre-order traversal sequence;

(g) 'beginning'; defined such that the 'next' operation gives the 'first';

(h) 'end'; defined such that the 'previous' operation gives the 'last'.

The basic operations of FTAM for access are to locate a FADU (by quoting a FADU identifier), erase it (eliminating an entire subtree, or, in the case of the root FADU, the entire file's contents), read it (in a great many ways, described in section 10.4.2.3), extend the data unit at its root node (provided the 'extend' operation is defined for the datatype of this data unit), or insert a complete new FADU, either as a sibling or as a descendant.

10.4.2 The nature of transferred material

10.4.2.1 The 'extend' operation

If a data unit at the root of a FADU is being extended, the transfer consists simply of a suitable datatype capable of being concatenated with the datatype currently present. Where a *document type* definition applies to the file, this concatenation operation will be fully defined in the document type. It is, however, possible to operate more generally. A data unit is defined as consisting of a series of one or more data elements, each data element being transferred as a single presentation data value. (A Presentation data value is a piece of information with a well-defined transfer syntax which is a *single* string of bits.) Where a document type is not used, every Presentation data value needed to transfer the data elements in the data units of the file's contents (including node names, which are a single Presentation data value) has to be defined as part of a single named abstract syntax. Where this approach is used, 'concatenation' of two series of data elements is implicitly defined as the composite series. Note that in the alternative case of *explicit* concatenation rules, specified in a document-type definition, the original data unit could be transferred as a single Presentation data value, and the result of extending could still be transferred as a single Presentation data value.

10.4.2.2 The 'insert' and 'replace' operations

In this case a complete FADU is being inserted (possibly with removal of an old one). FTAM provides a full specification of how to transfer the complete information content of an arbitrary FADU, with arbitrary data elements in its data units or used for its node names.

This transfer mechanism rests on an important Presentation (Layer 6) facility which recognises *instances of use* of named abstract syntaxes. One 'instance of use' of an FTAM-defined abstract syntax is reserved to carry what are essentially 'up', 'down' and 'node information' Presentation data values which cannot be confused with Presentation data values used to transfer node names or parts of data units.

Thus, there is no problem in supplying the full information needed to insert a new or replacement FADU. (Without this 'instance of use' mechanism in Layer 6, the FTAM protocol would have to define some form of escape mechanism to provide transparency for structuring information.)

10.4.2.3 The 'read' operation

The 'read' operation is the most complex of the operations on a file's contents, being moderated by the 'access context'. The access context allows the remote system reading the FADU to request any of the following collections of information:

(a) the complete information content of the FADU; in this case the transfer of Presentation data values is identical to that for an 'insert' operation;

(b) the data unit at the root of the FADU only; in this case the series of Presentation data values transferred is from the named abstract syntax(es) covering the data elements in the data unit;

(c) all data units in the FADU, concatenated by simply running together their series of data elements, taking them in pre-order traversal sequence (note that this may or may not make the original data unit boundaries undetectable);

(d) all data units in the FADU, but with their boundaries marked by presentation data values from an 'instance of use' of an FTAM-defined name abstract syntax;

(e) as (c), but data units for a particular *level* within the FADU only; (note that it is this concept, and this concept alone, which makes use of the idea of 'long arcs');

(f)   as (d), but data units for a particular level only;

(g)   the complete information content, *apart from* the data elements in the
      data units (the presence or absence of each data unit is, however,
      flagged).

10.4.3 The implementation problem

A similar implementation problem to the one discussed here occurs for
support of file attributes (see section 10.5, and particularly section 10.5.2)
and for concurrency control (see 10.6.4.2).

It is relatively easy to produce an implementation fully supporting the
general hierarchical structure and arbitrary abstract syntaxes for data units
and node names (using the bit patterns defined for transfer as the storage
form). Such an implementation, however, cannot support local access,
using standard utilities, to such files. The term 'separate' can be applied to
this approach.

On the other hand, mapping the simpler defined document types to local
structures, ignoring the more complex document types, and defining
vendor-specific document types for curious local file structures, is somewhat removed from the spirit of open interworking. Nonetheless, integration with local facilities is achieved, and this is called the 'integrated'
approach.

It seems likely that some middle position, possibly based on a data
dictionary approach, is the right one for implementors to take. The
requirement would be for local management to 'configure' the data
dictionary with local representations of all the Presentation data values
which are used to transfer the data units in a particular document type.
'Configure' would have to include specifying local storage mechanisms and
the form of displays on terminal, printer and plotter devices. Note that the
data dictionary would be used by all local utilities (*including the editor and,
for example, the FORTRAN run-time system*), so 'integration' is still
achieved, but in an open-ended way. This is called the 'configurable'
approach.

There was not sufficient understanding (in 1986) of the data dictionary
and the named abstract syntax *and* the document type concepts among
FTAM implementors (still less, widespread use of the concepts in local
editors) for the 'configurable' implementation to appear. Initial implementations of FTAM were either 'integrated' or 'separate'.

10.5 THE VIRTUAL FILESTORE

The virtual filestore consists of a collection of 'files'; a 'file' has *contents* and a number of *attributes* (one of which is the 'filename' attribute). In the simplest possible implementation allowed by the FTAM conformance statement, all attributes except 'filename' have fixed values, or are not supported. Thus, the only variable information which must be held for a file is its *filename* and its *contents*, corresponding to the capabilities of the simplest real filing systems.

At the other extreme, a 'full' implementation will hold values of the following 'file attributes', as additional information, with the file's contents:

Filename
Permitted actions
Access control
Storage account (the account to be charged for storing the file
Data and time of creation
Date and time of last modification (to the contents)
Date and time of last read access (to the contents)
Date and time of last attribute modification
Identity of creator
Identity of last modifier (of the contents)
Identity of last reader (of the contents)
Identity of last attribute modifier
File availability
Contents type
Encryption name
Filesize
Future filesize
Legal qualifications
Private use

The model of the file's contents is covered in sections 10.4 and 10.7. The following subclauses discuss those file attributes whose nature and range of values is not intuitively obvious from the name of the attribute.

10.5.1 Permitted actions attribute

The general model recognises the following possible actions on a file's contents or on its attributes or both:

read FADU
insert FADU
replace FADU

extend data unit at root of FADU
erase FADU
read attribute (an individual one)
change attribute (an individual one)
delete file (contents and attributes)

It also permits location of a FADU (see section 10.4.1 (a) to (h)) by:

traversal	use of 'first', 'next', 'begin', 'end' and 'last';
reverse traversal	use of 'last', 'previous' and 'end' included;
random order	FADU name, series of FADU names, integer, i.e. section 10.4.1 (a), (b), and (c).

When a file is created, the remote system states the actions and FADU location mechanisms it would like to be available. The system supporting the virtual filestore reduces this list to ones which it supports, and the result forms the file attribute. Only the actions and location mechanisms in the atrribute will be available for future work on the file. (Note, however, that in general a remote system accessing a file will not be aware of the value of this attribute. When an access is attempted, a so-called 'current permitted actions activity attribute' is established, known to both parties, constrained by this file attribute. The activity attribute constrains subsequent actions (see section 10.6 for further detail).)

Note also that the 'Contents type' file attribute (see below) may contain a reference to a specification which itself limits some of the possible actions.

This attribute cannot be changed after file creation.

This attribute should not be confused with access control (see below), which limits actions on a per-user basis, and can (subject to authorisation) be changed at any time. The permitted actions attribute is intended to reflect the capabilities of the medium and of the FTAM SASE supporting the virtual filestore. For example, a mapping of the virtual filestore on to a line printer would not permit 'read FADU'. A mapping on to a read-only device such as a digital image scanner or a card reader would permit *only* 'read FADU'. The flexibility for implementors to map the virtual filestore to any information-handling device is important for the widespread use of FTAM. It is capable of far more than simple support of a real disk filing system.

Note, however, that initial implementations of FTAM have concentrated almost exclusively on real disk filing systems.

10.5.2 Access control

This attribute is complex. It provides for access control based on any combination of:

(a) quoted passwords;

(b) an authenticated initiator identity; and

(c) an initiator location

independently applied to each of the actions listed in section 10.5.1 above. The attribute can be set up so that more than one password can be used interchangeably (so that a user quotes a different password for access from different locations, a different password is used for each of the actions listed in section 10.5.1, different users have different passwords, no additional passwords are needed for some authenticated initiator identities, and so on). The flexibility is high, with consequent implementation problems on unmodified filing systems in the 1980s. In many cases, real filing system mechanisms for access control to a real file are based only on an authenticated user identity, with little or no discrimination between actions ('read', 'write', 'delete', and 'change controls' are recognised in the more sophisticated systems). Sometimes a *single* additional 'file-password' is supported. Thus, it is easy to interpret access control setting in real systems in terms of a value of the FTAM access control attribute. It is not easy, however, to allow an initiator to set arbitrary FTAM values (i.e. provide full support for FTAM) and implement the controls by mapping on to the 'real' access controls enforced by the real filing system.

The way the FTAM functionality is formalised is to say that the access control attribute consists of a series of *conditions*, each potentially permitting any one or more of the possible actions. An action is permitted if any one or more conditions allows it.

Each condition consists of up to four components. The first is a Boolean vector with an element for each action. The second component is optional, and is a vector of passwords (an 'empty' element matches anything) one for each action. The user quotes a vector of passwords. If matches occur, the password vector generates a 'true', otherwise a 'false', and these values are 'anded' with the first component. The result is a set of actions (set A, say) which *might* be permitted by the condition. Finally, two further tests can *optionally* be added to the condition. The first is a test on whether the user identity (quoted and password-checked when access was established – see F-INITIALIZE in section 10.6.3) equals a given value. The second is a test on whether the Application entity title (see Chapter 3) quoted by the remote system making the access, equals a given value. If these components (if present) are satisfied, then the condition permits the actions in set A, defined above. If either is not satisfied, this condition does not contribute to the total set of permitted actions.

It should be noted that whilst the user identity is subject to authentication by checking a password, the Application entity title is open to masquerade by anybody with his own FTAM implementation. Thus the use of the 'initiator location' condition is often easily bypassed.

It should be clear to the reader that no real filing system of the 1980s provides precisely the FTAM form of access control. This problem again focuses attention on the question of 'separate' versus 'integrated' implementations of FTAM (see Section 10.4.3). Any 'integrated' implementation is only capable of accepting and enforcing a small subset of the possible values of the access control attribute. The 'configurable' approach is of no help here. Either the 'separate' approach has to be adopted, or a subset of FTAM has to be implemented, or the real file system has to be modified to fit the FTAM specification.

10.5.3 File availability

This attribute has two values – 'immediate' and 'deferred' – and can be changed by a remote user. *No semantics are specified.* Thus the following are all conforming implementations, 'fully' supporting the attribute:

(a) the value is stored but ignored; it expresses some (random) meaning of the remote user;

(b) the value expresses the remote user's wishes; if 'immediate', the file should be 'got ready' as soon as possible; if 'deferred', the file can be 'off-loaded'; the terms are implementation-dependent;

(c) 'immediate' means 'the disk is spinning', 'deferred' means 'the disk is dismounted';

(d) 'immediate' means 'on-site' ('spinning' or 'dismounted'), 'deferred' means 'in a bank vault ten miles away'.

This discussion shows that any attempt at further definition rapidly leads to the use of other highly implementation-dependent terms. For most systems there will probably be an obvious and useful definition to apply, which is then supported by the FTAM protocol and hence by FTAM initiator implementations. This is why presence of the attribute in the specification of FTAM is useful, even without semantic definition.

10.5.4 Contents type

This file attribute is established at file creation time, and cannot be changed. It holds a value identifying a file structure supported by the system supporting the virtual filestore. There are two forms this attribute can take. It is either:

(a) a single *document type* name; or

(b) a single *constraint set* name, with a single *abstract syntax* name.

Document types and constraint sets are discussed more fully in section 10.7.

A document type name references a definition which (within the general model) operates with complete freedom in determining the nature of the contents and the operations permissible on the contents.

A constraint set name references a restriction on the general structure and possible operations. The abstract syntax name identifies a set of data values which are available for node names and data units (wherever these are allowed by the constraint set).

The document type approach is in some senses less fundamental, but is more powerful and flexible than the alternative.

All operations on the file (inserting FADUs, extending data units, and so on) produce a 'contents' which satisfies the conditions imposed by this attribute.

The only mandatory conformance requirement in FTAM is to support at least one value of the contents-type attribute, of form (a) *or* (b).

The Standard defines a number of constraint sets (used in the definition of document types) and a number of document types (which include the definition and naming of some abstract syntaxes). Most early implementations support these document types.

10.5.5 Encryption name

This attribute is only weakly-specified in the initial FTAM Standard. It is really a placeholder for later standardisation. It is intended to provide an *aide-mémoire* to the remote user on the way the file was encrypted, and is ignored by the system supporting the virtual filestore.

10.5.6 Filesize and future filesize

The 'filesize' attribute, if supported, records the current size of the file. The 'future filesize' is always greater. The 'future filesize' can be changed by a change-attribute operation. The Standard explicitly states that, when writing to a file and exceeding the 'future filesize', the implementation can choose to:

(a) abort the transfer; or

(b) automatically increase the 'future filesize' value.

10.5.7 Legal qualifications

This is another weakly-specified attribute. The intent is to provide a place for holding any information required by national data protection legislation (e.g. the 'legal owner ' of the data). The UK Data Protection Act imposes no requirements which would use this attribute.

10.5.8 Private use

As its names implies, this attribute is not defined. Its use is 'strongly discouraged'. It requires (and provides a hook for) enterprise-specific standardisation.

10.6   FTAM SERVICE PRIMITIVES

The FTAM service primitives define the possible conceptual interactions of a remote user with the FTAM virtual filestore, via the FTAM SASE's use of the FTAM protocol. This section provides a description of the nature and sequence of the interactions, and the parameters of each interaction (These are defined in Part 3 of the FTAM Standard.)

The treatment in this text is an introduction and overview only, and the reader is referred to the FTAM Standard for details.

10.6.1 Regimes

FTAM operates through a series of nested 'regimes' (see illustration Fig. 10.3). For normal termination, an outer regime can only be exited if inner regimes have been exited, but the 'F-U-ABORT' primitive (an unconfirmed service) is available for the user to abnormally terminate all regimes, no matter how deeply nested. In addition, 'F-P-Abort' (a provider-generated service primitive) informs both users that the service provider cannot continue to provide the service, and that all regimes are lost. (This primitive generally results from an A-P-ABORT passed up by CASE, i.e. indirectly, from an N-RESET or N-DISCONNECT which is not recovered by the Transport Layer.) After exiting an 'inner' regime, the remote user can, in general, enter such a regime again (typically with different parameters).

The following regime structure is provided:

(a) The 'FTAM association regime'; this is the outermost regime; outside this regime, the two end-systems have no FTAM-specific knowledge of each other; on entry to the regime (*inter alia*), negotiation of FTAM capability occurs, accounts are established for charging, and the initiator can be identified; on normal exit, levied charges are returned;

(b) the 'selected regime'; this regime is entered when a specific file is referenced (an old one selected, or a new one created); concurrency controls (locks) can be applied at this time, and entry to the regime allows file attributes to be read and, where applicable, written, subject to access control; the file's contents are *not* accessible in this regime, except by entering an 'open regime' (see below);

(c) the 'open regime'; this regime makes the file's contents available for reading and/or writing; concurrency controls can be adjusted; within the regime a FADU can be located; it can then be erased, or a 'data transfer regime' (see below) can be entered to insert a sibling or descendant, extend a data unit, or read the FADU.

(d) the 'data transfer regime'; this is the innermost regime, and supports the transfer of actual data to read a FADU (in any access context – see section 10.4.2.3), insert a FADU (as sibling or descendant), or extend the root data unit of the FADU.

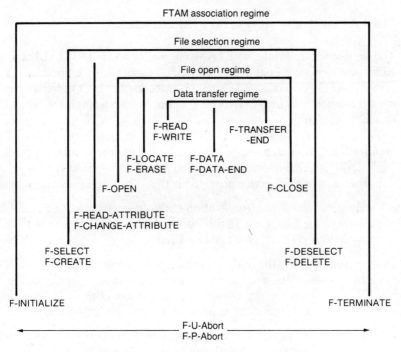

Fig. 10.3 FTAM regime nesting.

The names of the primitives used to enter and leave regimes are shown in Fig.10.3.

10.6.2 Parameters for exit from regimes

In general, a (normal) exit from a regime provides a 'result' parameter and (optionally) diagnostic information.

The 'result' parameter may indicate 'success', or a 'recoverable' (transient, e.g. due to 'congestion') error, or a 'non-recoverable' (permanent, i.e. to be resolved by human beings) error for the activity during the regime.

In addition, charges levied during the regime can be reported.

This 'recoverable'/'non-recoverable' distinction is much more important in FTAM than in lower layers that frequently make a similar classification of error codes. This is because in the so-called 'reliable service' (see section 10.8), FTAM provides for automatic, machine-performed recovery actions to prevent the (human) user from noticing errors flagged as 'recoverable'. 'Non-recoverable' errors are, however, immediately returned to the (human) FTAM user to sort out.

There are no other important parameters on regime termination.

10.6.3 Parameters for entry to the FTAM regime

The service for entering the FTAM regime (called F-INITIALIZE), is a confirmed service initiated by the remote user. It is embedded in A-ASSOCIATE (see section 9.1), which is embedded in P-CONNECT, which in turn is embedded in S-CONNECT. (Thus there is a single round-trip for the entire exchange.)

The end-system issuing the F-INITIALIZE request is called *The FTAM initiator*. The end-system supporting the virtual filestore (and responding to the F-INITIALIZE indication) is called *the FTAM responder*.

Parameters of this service provide for the following information flows:

(a) identification of the Application-entity-title assigned to the FTAM initiator (see Chapter 3.9); this information is carried using the 'CASE association control' protocol (see Chapter 9.1);

(b) a statement by the FTAM initiator of those features of FTAM ('functional units' and 'attribute groups') that it wishes to use in the FTAM association regime, with a response from the FTAM responder reducing this list to items supported by that implementation;

(c) agreement on use or non-use of the procedures designed to support a reliable service (see section 10.8);

(d) provision by the FTAM initiator of a user identification, account, and password known to the FTAM responder;

(e) agreement on use or non-use of the procedures of the commitment, concurrency, and recovery (CCR) Standard (see section 9.3); where these procedures are agreed, several of the FTAM service primitives carry parameters which can be used to start and end CCR atomic actions (these parameters are ignored in subsequent sections of this book);

(f) a 'contents type list' parameter (see below).

The 'contents type list' parameter needs some additional explanation. In the simplest form of Presentation Layer provision, all abstract syntax names to be used during a connection have to be declared when the connection is established. Connection establishment occurs when F-INITIALIZE is issued. Hence, it is necessary at this time (for the simplest use of the Presentation Layer) to identify all abstract syntaxes needed for any transfer which is to occur. The 'contents type list' parameter (whose form is that of a sequence of contents type file attribute values – see section 10.5.4) enables the FTAM SASE (assuming it has knowledge of any referenced document-type definition) to identify the abstract syntax names to be passed to the Presentation Layer for syntax negotiation in P-CONNECT.

If the 'context management functional unit' of the Presentation Layer is available on the connection, abstract syntax names can be declared for use at any time during a connection. In this case, the 'contents type list' parameter on F-INITIALIZE serves little useful purpose, save to identify early in the communication that the specified file structures are indeed recognised and supported by the FTAM responder.

It is important to note here that if context management is not in use, the remote user (the initiator) has to know what document types he is going to be working with *before* selecting files and reading their content's type attributes. Context management is desirable in order to support friendly and flexible interfaces for human users, although early initiator implementations of FTAM (supporting very few document types) attempt to negotiate the availability of all the ones they support.

10.6.4 Parameters for entry to the 'selected' regime

The 'selected' regime is entered either by selecting a file or by creating a new one. In both cases the major parameter is the file name; in the case of 'create', additional parameters specify the initial file attributes of the newly-created file. (An option exists to allow the FTAM initiator to request 'create me a new file if one doesn't exist with the given name, or empty and re-use the file otherwise'.)

The services (called F-SELECT and F-CREATE) are both confirmed services, issued by the FTAM initiator. Other parameters on entry to this regime are available to:

(a) determine the possible actions on the file ('requested access' and 'access passwords' parameters);

(b) determine the concurrency controls to be applied to the file; and

(c) nominate an 'account' to be charged for activity in the regime (temporarily overriding that passed on F-INITIALIZE).

10.6.4.1 Requested access

The 'requested access' parameter is a Boolean vector with elements corresponding to the actions listed in section 10.5.1. The 'access passwords' are used to match those in the conditions of the 'access control' attribute for the file (see section 10.5.2). The resulting access which is allowed is the least of what was requested, what the 'permitted actions' allows, and what the 'access control' allows. This establishes the 'current permitted actions' activity attribute, whose value is returned on the response primitive.

10.6.4.2 Concurrency control

The concurrency control parameter also needs some discussion. This is present in both the Request/Indication and in the Response/Confirm. It is a vector, with one element for each of the actions in section 10.5.1. Each element takes one of four values (relating to the duration of the selected regime) saying:

(a) I want to perform the action, and others are also allowed to; or

(b) I want to perform the action, but please prevent others from doing it; or

(c) I will not do the action, but others are allowed to do it; or

(d) I will not do the action, and please prevent others from doing it.

These options become clearer when considered against combinations of specific actions such as insert a FADU, erase or replace a FADU, read, or delete the file.

There is clearly a potential interaction with the results of 'requested access' processing (see section 10.4.1). For the purpose of concurrency control, failure to obtain the right to perform an action clearly eliminates the statement in the concurrency control parameter that 'I want to do it', but you are still allowed to prohibit others from doing it while you have access. (Thus, for example, you may not have the right to delete a file, but you still want to prevent its deletion while you are reading it.)

On the request/indication, the FTAM initiator states the concurrency controls he wants. On the response/confirm he is told what he has got.

If the initiator offers to allow others to perform an action, the FTAM responder may none the less apply concurrency controls locking others out. If the initiator says he does not want to perform an action, the FTAM responder may nonetheless assume he might and hence delay requests for exclusive access by other users. Resulting concurrency controls are returned in the response/confirm.

This right to vary the controls is vital. No 1980s computer system implements, in its real filestore, the granularity of concurrency controls described above. It cannot be implemented in full (except using the 'separate' approach – see section 10.4.3). The above variations, however, permit every request to be converted into one for 'shared reading' or into one for 'exclusive write/delete'. These types of concurrency control are very common in real systems.

It is important here to note one major weakness of the FTAM specification. A decision was taken early in FTAM development that no constraints would be placed on how quickly a response was generated. Thus, it is *an implementation option* whether, on failure to obtain the locks needed for concurrency control, an immediate (recoverable error) response is given, or whether the FTAM responder waits (perhaps indefinitely) until the locks can be obtained. Systems initiating an F-SELECT or F-CREATE (or an F-OPEN) need to be capable of dealing with either situation (by a repeated retry on a recoverable error, or by an F-U-ABORT on an excessive wait).

10.6.5 Parameters of read and change attribute

The read and change attribute services (F-READ-ATTRIBUTE and F-CHANGE-ATTRIBUTE) are confirmed services, initiated by the FTAM initiator.

The parameterisation allows one or more attributes to be read or changed on each use of the service. On reading, a value may be obtained, or an 'unset' value, or a 'no value available' response. The latter means that whilst the FTAM responder understands and can 'talk about' this attribute, it actually does not record values. The conformance statements allow most file attributes to be supported at the level of: 'I can talk about it, but will always return no-value-available'. This is an important aid to implementability.

A word is needed here about when a change-attribute *really* takes effect.

In particular, if there is a 'crash' *after* the change-attribute response/confirm, but *before* the selected regime is exited, can the FTAM initiator guarantee that the change has occurred? The answer is, 'NO'. It is implementation-dependent whether 'commitment' to the change is delayed until the selected regime is exited normally.

A similar question (and a similar answer) arises for the scenario of one FTAM initiator changing an attribute and a different one reading it immediately afterwards.

10.6.6 Parameters for entry to the open regime

The open regime (giving access to the file's contents) is entered by F-OPEN. This is a confirmed service initiated by the FTAM initiator.

Parameters of this service provide for:

(a) variation of the actions which are to be allowed (see processing mode, below);

(b) establishment of the contents type value to be used (see section 10.6.6.2 below); and

(c) variation of the concurrency controls on the file (see concurrency control variations below).

10.6.6.1 Processing mode

A parameter called 'processing mode' (a somewhat unhelpful name) allows the set of actions which were established by processing the 'requested access' parameter on F-SELECT or F-CREATE to be reduced for the duration of this open regime.

Some actions which are allowed by the 'access control' and 'permitted actions' file attributes (the only constraints on 'requested access' on F-SELECT) may none the less be forbidden by rules associated with the 'contents-type' value. On F-OPEN, the 'processing mode' (a mandatory parameter) has to request only operations permitted by the 'contents-type' value which is applicable (see below), otherwise the open fails (non-recoverable error).

10.6.6.2 Contents type

When the open regime is established, agreement is achieved on the nature of the file type to be accessed.

In the simplest case, the request/indication is null, or carries a value equal to that in the 'contents type' file attribute. In this case, the value in the 'contents type' file attribute is returned. If the Presentation context management functional unit is not available, all necessary abstract syntaxes must already be agreed for use. If it *is* available, the *responder* takes the initiative (at the same time as he issues the F-OPEN response PDU as P-DATA) to issue a P-CHANGE-CONTEXT primitive to establish the abstract syntax. The F-OPEN confirm is not issued until agreement on all necessary abstract syntaxes has been reached.

In the more general case, the value in the request/indication is 'simpler than' the value in the 'contents type' file attribute, and is returned unchanged. 'Simpler than' involves a fairly complex set of rules, together with 'simplifications' defined for each document type (see, for example, section 10.7). In this case, 'processing mode' is not allowed to permit modification of the contents and the contents look, for all purposes, as if they were the 'simpler' file type.

The 'simpler than' mechanisms allow simple initiator implementations to read (but not write) complex (indexed, for example) files as if they were a simple unstructured data unit at the root node.

10.6.6.3 Concurrency control variations

The concurrency control established when the selected regime was entered can be 'tightened'. This means that, for any action, the statement 'I will not do it' can be changed to 'I want to do it', and 'Others may do it' can be changed to 'Others may not do it'.

It is worth noting why 'tightening' is the only allowed variation. FTAM attempts to ensure that *exiting* a regime is always possible. If F-OPEN could *relax* a concurrency control, exit from the open regime (by F-CLOSE) would have to reinstate it, and this could be impossible within a reasonable timescale. Thus only 'tightening' is allowed on F-OPEN.

10.6.7 Parameters for F-ERASE and F-LOCATE

The F-ERASE and F-LOCATE services are confirmed services available in an open regime. The only important parameter is the FADU identifier, discussed in section 10.4.

10.6.8 Parameters for entry to the data transfer regime

When data transfer (to or from the virtual filestore) is initiated, parameters are available to:

(a) identify the FADU affected (see Section 10.4); and

(b) indicate
  (i) read in some access context; or
  (ii) extend, insert, or replace.

Within the data transfer regime, the data to be transferred is shipped using Presentation Layer services. One or more so-called F-DATA primitives are mapped to one or more so-called P-DATA primitives, ending with F-DATA-END, mapping to a special presentation data value of P-DATA, which is in the same presentation context as the FTAM PDUs (and, hence, is distinguishable from any data value in a data unit – see section 13.4). In the extreme case, an implementor can transfer the entire information, including the F-DATA-END termination, as a series of presentation data values on a single P-DATA.

The so-called 'user-correctable service' and corresponding protocol include facilities for inserting checkpoints for recovery (and recovery primitives). The way these are to be used is, however, specified only for the so-called 'reliable service' (see section 10.8). Use of these functions in the basic 'user-correctable service' requires additional enterprise-specific standardisation. They can be implemented (for general-purpose implementatiuons) only as a part of support for the 'reliable service', and are discussed in that section.

10.7   CONSTRAINT SETS AND DOCUMENT TYPES

'Constraint set' definitions can be seen as one step on the way towards the 'document-type' definitions that reference them. They are, however, as has been seen earlier (see section 13.5.4), allowed to be used directly in the 'contents-type' file attribute and parameters. (For direct use, however, an abstract syntax name is also needed, and these are only defined in FTAM as part of document-type definition.)

'Constraint set' definitions are concerned entirely with restricting the *structure* of a file. For example, the simplest constraint set requires the file to consist only of a root node, with no node name, and with a data unit (of unspecified type) at the root node.

'Document type' definitions go further, and define not only constraints on the structure (usually by reference to a constraint set), but also the datatypes forming the data units, operations such as 'concatenation' of data units or 'extension' of data units, and 'simplifications' (see section 10.6). Document type definition is thus a more powerful and flexible tool for completely defining file structures.

The following subclauses discuss some of the constraint sets and document types whose definitions are expected (at June 1986) to appear in the FTAM Standard.

10.7.1 Constraint sets

Seven constraint sets are defined. The first three are fairly easily mapped to real systems of the 1980s, and are the most important for early implementations.

10.7.1.1 'Unstructured' constraint set

This constraint set restricts the general model to a root node (and no further node) with a data unit (possibly empty). Allowed actions are read, replace, extend the data unit, and erase (replace the data unit with an empty one).

This fits fairly well with the ability of many operating systems to read, write, and add to the end of a file. Where a particular implementation can't support 'add to the end of' or 'erase', the extend and erase operations can be prohibited by eliminating them from the 'permitted actions' attribute on file creation.

10.7.1.2 'Sequential flat' constraint set

This constraint set restricts the general model to a root node without a data unit, and Level 1 leaf nodes each with a (possibly empty) data unit.

There are no node names. A FADU can only be located for reading by its position in the pre-order traversal sequence. 'Erase' is only available on root FADU to empty the file. 'Insert' is only possible at the 'end' of the file. 'Extend' is not allowed, nor is 'replace'. Thus for this constraint set, once information (in the form of 'records', for example) has been added (at the end only), it cannot be modified (replaced or extended) in any way, or deleted (save by deleting the whole file's contents), but random access to the individual 'records' for reading (by their position in the sequence) is possible. The entire file can be read with or without the boundaries between the 'records' being suppressed (important only where the 'record' is a series of data values, rather than a single data value).

10.7.1.3 'Ordered flat' constraint set

This constraint set differs from that described above in that it allows node names, which can be used for 'locate', and in the ability to insert new FADUs within the Level 1 sequence, and in the ability to erase any FADU.

The node names are used to define the order of the 'records', independently of the order of insertion. Thus, they can be used to hold 'keys' to the records. The type of node name, and the algorithm for ordering them, is not defined in the constraint set. (Thus, this constraint set name cannot be used in the contents-type attribute or parameter, as additional information is needed which cannot be supplied by an abstract syntax definition.)

10.7.1.4 'Ordered flat, unique names' constraint set

This is an ordered flat constraint set with the sole added constraint that node names must all be distinct. ('Insert' fails if a name is duplicated.)

10.7.1.5 'Ordered hierarchical' constraint set

The only constraints imposed by this constraint set are on the nature of node names.

Node names of siblings are required to be distinct, and siblings are ordered by an ordering defined for the node names. Again, the type of node names and ordering algorithms require to be separately specified, typically in a document type definition, before this constraint set can be implemented.

10.7.1.6 'General hierarchical' constraint set

The only constraint imposed on the general model by this constraint set is that all node names at a given level shall be 'the same data type'.

10.7.1.7 'General hierarchical with unique names' constraint set

This is a general hierarchical constraint set except that the names of siblings are required to be distinct.

10.7.2 Document types

Four document types are defined. The first is generally supported by all implementations.

10.7.2.1 *'Unstructured text' document type*

A file described by this document type name satisfies the 'unstructured' constraint set, with the single data unit consisting of zero or more instances of a 'GraphicString' data value. A 'GraphicString' is defined as an unbounded sequence of printing characters (plus 'space') *from any character set registered in the ISO Register of Character Sets.*

No 1980s operating system can support all values of this document type (using the 'integrated' approach), and consequently implementors frequently subset to the ASCII character set, the UK character set (subtly different) and so on.

Ideally, separate document types would be defined for the character sets of all languages and combinations of language, but this would be an awesome task. This problem has led to proposals for *parameterisation* of document types, but this approach has not been taken for the initial OSI Standards. (Note on proof reading – parameterisation is now included.)

10.7.2.2 *'Binary' document types*

Two document types are defined for 'binary' files. One references the 'unstructured' constraint set and the other the 'sequential flat' constraint set. In both, the data unit consists of a sequence of OctetString data values. (Note, OctetString, not BitString.) In the first, however, there are no visible boundaries in the sequence, and in the second there are.

The first of these document types is easy to implement on most systems, but unless transfers are being made between like machines, the facility is of use only with added definition of the meaning of the binary, or for a relatively sophisticated user's purposes.

10.7.2.3 *'Sequential text' document type*

This document type references the 'sequential flat' constraint set.

Each data unit consists of zero or more data values, each of which is a 'GraphicString'.

A 'simplification' is defined for this document type, resulting in the 'unstructured text' document type by concatenating all data units (and ignoring their boundaries). Thus, any file with this value in its contents type file attribute can always be read (but not written) by quoting in the F-OPEN a contents type parameter of 'unstructured text'.

10.7.2.4 'Simple hierarchical' document type

This document type is *not* a complete specification, and is not usable in the content's type parameter or attribute.

It is intended to be referenced by other document type definitions (which do not exist in the initial Standard).

It references the 'general hierarchical, with unique names' constraint set, and does no more than define the form of node names. Each node name is defined to be a single 'GraphicString'.

10.7.2.5 Implementation

A number of high-profile and large 'closed' groups will define additional document types. Vendor-specific types will also be defined.

In terms of what is fully implementable without additional specification, the only values possible for the contents type attribute are document type names of:

(a) unstructured text (subsetting GraphicString);

(b) binary (both forms);

(c) sequential text (subsetting GraphicString).

The real challenge FTAM poses for implementors is to design real systems (operating systems, file storage, editors, run-time systems, and protocol handlers) which can flexibly handle any document type by appropriate configuration using some form of data dictionary.

10.8   THE RELIABLE SERVICE

In the simplest implementations of FTAM, the 'user-correctable service', a lower layer 'Disconnect' (possibly generated by failure of the remote end-system) produces an F-P-ABORT which terminates the FTAM activity. All regimes are abnormally terminated, and in particular, all concurrency controls are released.

If such a failure occurs during an open regime, the FTAM Standard states that the content of the file is implementation-dependent. This (together with the release of concurrency controls) makes any attempt at automatic recovery by the FTAM initiator very difficult (particularly after an 'append'); in general, such automatic recovery cannot be implemented; any 'failure' is reported as an error to the (human) user, who must use human intelligence to determine how much of what he wanted done has been done, and how much must be attempted again.

When the FTAM regime is established, it is, however, possible to negotiate and agree use of procedures supporting the 'reliable service'. These procedures are applied only during an open regime. The remainder of this section describes the operation of FTAM when these procedures are applied.

The procedures apply first to the FTAM initiator, and involve two extra parameters on F-Open. The first and most important extra parameter is an *activity identifier* which identifies this piece of reliable FTAM activity for this initiator/responder pair. The second parameter agrees (by negotiation) whether failures can be recovered by continuing from a checkpoint (inserted during the data transfer regime by an F-CHECK primitive mapping to a P-SYNC-MINOR) or can only be recovered from the start of the latest data transfer regime. A third option exists which allows use of recovery procedures to be 'switched off' for the duration of the open regime.

Before issuing an F-OPEN request (which, in the reliable service, may cause concurrency controls to be retained following a failure), the FTAM initiator records that he has responsibilities to the FTAM responder, and in particular records the activity identifier he has generated. This record takes the form of data which is protected from crashes – typically, it needs to be written to disk and the file closed; it is called a *docket*.

On receipt of the F-OPEN indication and before issuing the F-OPEN response, the FTAM responder writes up its own docket, containing the activity ID, an identification of the selected file, and the concurrency controls in operation. The docket is used after a crash to reinstate concurrency controls.

If crashes occur *after* issue of an F-OPEN request but *before* a docket is written up by the FTAM responder, the FTAM initiator will (in accordance with the rules related to its docket), attempt to recover the activity by issuing a new F-INITIALIZE followed by an *F-RECOVER* primitive quoting the same activity identifier. The FTAM responder will reply saying 'activity identifier unknown' (because the crash occurred before its docket was written), and this is treated as a fatal error by the FTAM initiator; recovery is not possible because the open regime was never really entered.

A similar situation occurs at the end of the open regime, where the FTAM responder deletes its docket first (before issuing the F-CLOSE response), and the FTAM initiator deletes its docket on receipt of the F-CLOSE confirm. (Note, however, that the transfer has been ended, and its success or failure signalled, by the confirmed F-TRANSFER-END service. F-CLOSE merely tidies up the dockets.)

Apart from these 'transient' situations (in both of which the state of the file's contents is fully predictable to the FTAM initiator), failures within the open regime cause the FTAM responder to wait (with concurrency controls in place) for the initiator to attempt recovery. The FTAM initiator repeatedly attempts to issue an F-INITIALIZE followed (if successful) by an F-RECOVER, ignoring any 'transient' failures.

The F-RECOVER specifies (in addition to the activity identifier) the number (within the open regime) of the data transfer to be recovered. This number guards against crashes at the end of a data transfer regime which could leave the two ends in doubt about which was the 'last' (current) data transfer.

Finally, if recovery *within* a data transfer (using checkpointing) was negotiated on the F-OPEN, the F-RECOVER is followed by an F-RESTART primitive which negotiates the checkpoint from which the data transfer is to continue. Otherwise, F-RECOVER is followed by F-READ or F-WRITE.

The reader who has mastered the reliable protocol analysis techniques of section 9.3.11 will be anxious to apply them to the FTAM protocol. For this he is referred to the FTAM Standard: Part 4. In terms of the 'rules of thumb' at the end of section 9.3.11, we should note the existence of F-TRANSFER-END; this is a crucial part of the protocol which provides the double-handshake and allows FTAM to get a 'yes, yes, no' response to the questions of section 9.3.11.

Nonetheless, the reader is strongly recommended to practise his skills with a detailed analysis of the FTAM reliable protocol.

10.8.1 Scope of the reliable service

The reliable service provides protection for changes to file contents, enabling changes (such as extensions of a data unit) to be reliably achieved.

No such mechanism exists for changes to attributes, file creation, or file deletion.

10.8.2 Implementability

Incorrect implementations of the reliable service can leave FTAM responders with concurrency controls in place on a file, whilst the initiator believes his procedures are complete and makes no attempt to recover.

Incorrect implementations can arise for several reasons. One is misunderstanding the requirements of the FTAM Standard. Another is misunderstanding the local operating system. A docket should be created by an FTAM initiator before issuing an F-OPEN request. This means, typically, that information has to be written to a disk file, the file closed, *and all relevant operating system buffers written up*, before the protocol message is sent. This can be difficult to ensure on some systems, so a crash following the issue of F-OPEN can result in a docket and concurrency controls at the responder's end, but no docket and no recovery attempt at the initiator's end.

Another problem arises because of a known 'bug' in the FTAM protocol Standard. Suppose an initiating implementation, during an open regime, decides to kill the whole activity by issuing an F-U-ABORT service primitive (which it is allowed to do, even in the reliable service). F-U-ABORT is an unconfirmed service, losing the connection, and can 'collide' with an up-coming loss of connection from the lower layers. Thus, the FTAM initiator sees the reliable procedures as complete and takes no further actions (because it issued F-U-ABORT). The FTAM responder sees only the communications failure, and keeps concurrency controls in place, awaiting restart.

The FTAM Standard documents this possibility and recommends FTAM initiators not to do this (relatively easy advice to follow), and warning FTAM responders that it might happen (more difficult to take account of).

For the above reasons, FTAM responder implementations are likely to choose to release concurrency controls and delete dockets if initiator implementations are too slow about restarting. There is, however, (due to the fundamental approach to timers described in section 10) no suggestion in the FTAM Standard, nor any parameter in any service primitive, to help choose a suitable time for retention of controls.

10.8.3 Use of the CCR Standard

The CCR Standard (see section 9.2) includes its own procedures for concurrency control and recovery, as well as for commitment.

When use of these procedures with FTAM is negotiated (on F-INITIALIZE), the CCR procedures take precedence and reduce some of the problems identified in section 10.8.2 above, as well as providing recovery mechanisms which can encompass a change attribute service and file deletion/creation.

Few early implementations, however, support CCR in FTAM.

10.9 TRANSFER OF FTAM DOCUMENTS

This section describes the way in which text documents are transferred using FTAM. It is a detailed discussion assuming knowledge of ASN.1, and can be passed over on a first reading of this text.

10.9.1   Presentation service review

First, let us review the material on Layer 6 presented in Chapter 8, in relation to FTAM.

The Presentation service conveys, in each P-Data primitive, *zero, one or more* pieces of information called *presentation data values.*

A presentation data value can be transferred by the Presentation service if and only if there is at least one defined encoding for the presentation data value, which represents it by *a single undivided string of bits.* The string of bits is not required to be self-delimiting, nor is it required to be an integral number of octets.

It is often convenient, when defining (in an Application Layer standard) information to be transferred, to group the information together into one or more ASN.1 datatype definitions. These ASN.1 datatype definitions are used to define formally the information content of each Presentation data value which the Application requires to be transferred. Frequently, *but not necessarily*, there is a one-to-one correspondence between values of ASN.1 data types and Presentation data values. Whether ASN.1 is used or not, the Application protocol has to clearly identify all the Presentation data values which are to be transferred.

The Presentation data values to be transferred during some application context are collected together into one or more groups of Presentation data values. Each of these groups of Presentation data values forms what is called a *named abstract syntax* and is assigned a name (of ASN.1 type Object Identifier).

Where values of ASN.1 datatypes are in one-to-one correspondence with the Presentation data values in a named abstract syntax, there will often (but not necessarily) be a single ASN.1 module definition corresponding to the named abstract syntax.

Associated with each named abstract syntax there is one or more *transfer syntax names.*

For each transfer syntax name there is a specification of a representation (using a single undivided string of bits) for every Presentation data value in the named abstract syntax with which the transfer syntax name is associated. The string of bits representing a particular Presentation data value is called a transfer syntax for that Presentation data value and is distinguished from other possible representations of the same Presentation data value by the transfer syntax name.

The same string of bits can represent different Presentation data values if either:

(a) the Presentation data values are from different named abstract syntaxes; or

(b) the Presentation data values are from the same named abstract syntax, but are being encoded according to different transfer syntax specifications.

A transfer syntax specification is well-formed if and only if, for all Presentation data values in the named abstract syntax, each Presentation data value has a different representation. An Application protocol designer should normally ensure his transfer syntax specifications are well-formed. If he does not, the interpretation (by a receiver) of Presentation protocol encoding carrying a Presentation data value will be context-sensitive, or may even be ambiguous.

The Presentation service is only capable of transferring Presentation data values which are part of some named abstract syntax, and requires at least one associated transfer syntax name and specification.

It is common practice to use a collection of ASN.1 datatypes (often grouped into an ASN.1 module) to identify the Presentation data values in a named abstract syntax, and to define each Presentation data value as the value of a single ASN.1 datatype. Reference to the bit patterns produced by the ASN.1 Basic Encoding Rules (when applied to values of the datatypes) provides a well-formed transfer syntax specification for the named abstract syntax, provided tagging is used 'appropriately'. 'Appropriate' use of tagging is ensured by the rules of ASN.1 provided every datatype to be used for a Presentation data value is listed in a top-level ASN.1 'Choice' construct, and any tagging in the Choice is included in the data type used for the presentation data value. This is a sufficient (but not a necessary) condition for a transfer syntax to be well-formed.

It can, however, sometimes be convenient to define a single Presentation data value as a series of ASN.1 datatype values. (*This is done in FTAM document type definitions.*) In this case, the transfer syntax specification for such Presentation data values needs to be *more* than simply a reference to the ASN.1 Basic Encoding Rules, as these are only applicable to encoding the value of a *single* ASN.1 datatype.

A common piece of additional specification is simply to require concatenation of the bits produced by the ASN.1 Basic Encoding Rules, in order to specify the transfer syntax for this more complex Presentation data value. Note that the ASN.1 Basic Encoding Rules produce self-delimiting encodings, so simple concatenation does not lose information. The resulting transfer syntax for these Presentation data values, defined in this way, is *not*, however, a self-delimiting encoding.

To summarise the requirements of the Presentation service; any Application Layer protocol needs to:

(a) define clearly the Presentation data values to be transferred on each Presentation service primitive;

(b) to group these into appropriate groupings and assign an abstract syntax name to the group; and

(c) to define and name one or more transfer syntaxes for the abstract syntax.

These requirements are broader than, and independent of, ASN.1. When ASN.1 is used, ASN.1 datatype definitions can be used in the definition of Presention data values, singly or combined. A grouping of ASN.1 datatypes into ASN.1 modules is available, but is logically independent of the grouping of Presentation data values into named abstract syntaxes. The ASN.1 Basic Encoding Rules can be used in the specification of a transfer syntax for Presentation data values defined using ASN.1 datatypes.

10.9.2 The FTAM 'unstructured text file' document type

The FTAM–1 document type (and others), exploits the facility of having more than one ASN.1 datatype value in each Presentation data value.

The document is defined as:

'an unbounded sequence of character strings (lines of text). Each character string (line of text) is unbounded, and contains characters from any of the graphics character sets (plus space) registered in the ISO Register of Character Sets.'

The abstract syntax is defined so that each Presentation data value in the named abstract syntax is: 'an indefinite series of instances of the ASN.1 data type GraphicString' (each carrying one line from the document) and the transfer syntax is defined as: 'applying the ASN.1 Basic Encoding Rules to each ASN.1 datatype in the (Presentation) data value and concatenating the resulting octets'.

The document is intended to be transferred by grouping the lines of text into one or more Presentation data values, carried on one or more P-Data primitives.

The number of lines of text (from one to the entire document) to place in each Presentation data value is a sending implementation option. So is the use of a single P-DATA primitive to carry all the resulting Presentation data values, or the use of many. In the extreme case there could be one line of text in each Presentation data value and one Presentation data value in each P-DATA primitive. It is important to note here, however, that *none of the lower layers concatenate P-Data primitives*. This one line per P-DATA means one message across the network per line. Thus, it is sensible to either put several lines in each Presentation data value, or to put several Presentation data values on each P-DATA.

The transfer syntax for a Presentation data value which is chosen to be a single line of text is the bits defined by the ASN.1 Basic Encoding Rules for a GraphicString. The transfer syntax for other Presentation data values is the concatenation of such bit strings. As the receiver does not know how many such strings were put in each presentation data value, the transfer syntax for such Presentation data values is not self-delimiting, and relies on the Presentation layer to delimit it. (This is a service the Presentation Layer provides.)

10.9.3 Encoding of the P-Data primitive

The PDU carrying the P-Data semantics is defined in the Presentation protocol as:

SEQUENCE OF EXTERNAL

Each EXTERNAL carries a single Presentation data value from the P-Data primitive. The EXTERNAL identifies the abstract and transfer syntax for the Presentation data value (by carrying the integer value identifying the negotiated Presentation context), and carries the defined encoding. Where the Presentation data value is the value of a single ASN.1 datatype, it can be carried in the EXTERNAL as ASN.1 type 'Any' (no additional encoding); where it is a *series* of ASN-1 datatype values, it is carried as type Octetstring (or, as a sender's option, as type Bitstring).

Figure 10.4 shows the process of producing three P-DATA primitives for a nine-line document, and the following text describes the encodings at each of the stages 'A' to 'G'. Note that the choice of groupings into Presentation data values and into P-DATA primitives is a sender's option chosen here purely to illustrate various cases.

The following text mentions the number of octets overhead per line, and the percentage overhead. These figures assume ten octets overhead per S-DATA for PC1 for Session, Transport, and all the lower layers. The figure is typical. The percentages given assume an average of twenty characters per line. Putting the entire document in a single Presentation data value produces an average of about 2 octets overhead per line, or about 10%.

Let us now describe the coding at points 'A' to 'G'.

'A'   Each character string goes into an ASN.1 datatype called GraphicString. If all the characters are from the International Reference Version of ISO 646 (note, this does *not* include £ or $), *a single octet encodes each character*. For other characters, escape sequences will be needed in each encoding to invoke character sets including the characters required. For example, 'A$B' could be encoded by the octets:

4/1 ESC 2/8 4/2 2/4 ESC 2/8 4/0 4/2
A (Select ASCII) $ (Select ISO646-IRV) B

It is important here to note that the letters A to Z (and other

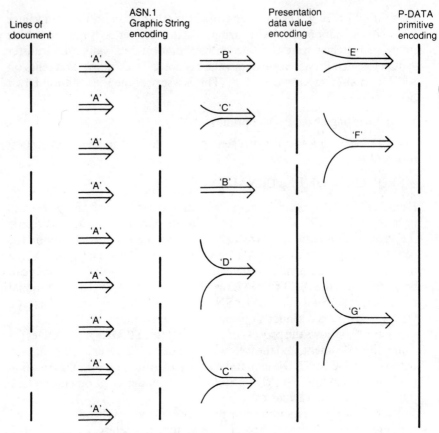

Lines of
document

ASN.1
Graphic String
encoding

Presentation
data value
encoding

P-DATA
primitive
encoding

Fig. 10.4 Example of transfer of a document.

characters) appear in many registered character sets. A sender would normally transmit characters in the ASN.1 default character set if they appear there, but may not always do so if (as in the case above), he has already escaped into another character set which also contains the character. Thus a receiver should be prepared to recognise these characters even if an escape into ASCII, or into the UK character set (both of which are registered) is made. A receiver also has to decide how to handle characters which are not in his own character set. The ideal would be to define a local escape mechanism for representing such characters. An alternative would be a mapping, with an FTAM warning diagnostic. A third alternative – appropriate if large parts of the document are unintelligible – is to abort the transfer. Decisions in these areas should be carefully documented for users of the implementation.

It should be noted that ASN.1 specifies reversion to the

International Reference Version of ISO 646 at the start of each GraphicString.

'B'   The Presentation data value is encoded as:

ASN.1 External.

We get, therefore

T   (Universal 8) – EXTERNAL
L   Any ASN.1 option

V
{
T   (Universal 2) – INTEGER
L   Any ASN.1 option
V   Presentation context identifier for the
      Presentation data value's context
T   (0) – because it is a single ASN.1 datatype
L   Any ASN.1 option

    V {
    T   (Universal 25) – GraphicString
    L   Any ASN.1 option
    V   The character encodings
         listed under 'A' above.

Note: Nine octets overhead for one line of text.

'C'   The presentation data value contains two lines of characters. We get, therefore:

T (Universal 8) – EXTERNAL
L Any ASN.1 option

V
{
T   (Universal 2) – INTEGER
L   Any ASN.1 option
V   Presentation context identifier
      for the Presentation data value's
      context
T   (1) – not ASN.1, but octet-aligned
L   Any ASN.1 option

    V {
    T   (Universal 25) – GraphicString
    L   Any ASN.1 option
    V   Encodings 'A'
    T   (Universal 25) – GraphicString
    L   Any ASN.1 option
    V   Encodings 'A'

Note: Eleven octets overhead for two lines of text.

'D'   The Presentation data value contains three lines of characters. We get, therefore:

T   (Universal 8) – EXTERNAL
L   Any ASN.1 option

    T   (Universal 2) – INTEGER
    L   Any ASN.1 option
    V   Presentation context identifier
        for the Presentation data value
        context
    T   (1) – Not ASN.1, but octet-aligned
    L   Any ASN.1 Option

V

        T   (Universal 25) – GraphicString
        L   Any ASN.1 option
        V   Encodings 'A'
        T   (Universal 25) – GraphicString
    V L   Any ASN.1 option
        V   Encodings 'A'
        T   (Universal 25) – GraphicString
        L   Any ASN.1 option
        V   Encodings 'A'

Note: Thirteen octets overhead for three lines of text. (Reduces to two + per line if the entire document is a single Presentation data value.)

'E'   The P-DATA carries a single Presentation data value (which is a single line of the document). We get, for the entire octets of the S-DATA primitive user data:

    T   (Application 1) multiple-context-data
    L   Any ASN.1 option
    V   Encoding 'B'

Note: Twenty-one octets overhead for one line of text (including lower layers), an overhead of about 100% (assuming twenty-character lines).

'F'   The P-DATA carries two data values (constituting three lines of the document). We get, for the octets of the S-DATA primitive user data:

    T   (Application 1) multiple-context-data
    L   Any ASN.1 option
    V   Encoding 'C'
       Encoding 'B'

Note: Thirty-two octets overhead for three lines of text (including lower layers), an overhead of about 50% (assuming twenty characters per line).

'G' The P-DATA carries two data values, constituting five lines of the document. We get, for the entire contents of the S-DATA primitive user data:

   T (Application 1) multiple-context data
   L Any ASN.1 option
   V Encoding 'D'
      Encoding 'C'

Note: Thirty-six octets overhead for five lines of text (including lower layers), an overhead of 36% (assuming twenty characters per line). A greater level of concatenation would approach a figure of about 10% for the overhead.

10.9.4 Other Presentation data values

Some of the more complex FTAM document types use Presentation data values in a document-specific named abstract syntax, but also make use of Presentation data values from the main FTAM-named abstract syntax. Although these pieces of information are both carried by ASN.1 datatypes, they cannot be simply concatenated into a single Presentation data value because they are from different named abstract syntaxes. They are carried as separate data values, but can still appear on the same P-DATA primitives with parts of the document text.

Similarly, the document is terminated by an FTAM PDU carrying the F-DATA-END semantics. This is again carried by a Presentation data value from the main FTAM abstract syntax, but can be carried on the same P-DATA as the document itself.

FTAM provides for 'concatenation' of the PDUS used in a BEGIN-GROUP to END-GROUP grouping. Note, however, that this is done at the level of putting several Presentation data values (compulsorily) into the same P-DATA. The one-to-one correspondences between PDUs, ASN.1 datatypes, and Presentation data values is maintained.

10.10 TYPICAL FTAM IMPLEMENTATIONS

Some of the implementation problems and options have been discussed earlier in this chapter. Those discussed have concentrated largely on the problems an FTAM responder faces in mapping the virtual filestore on to a real system, and in protecting against failure of an initiator to institute recovery when concurrency controls are in place.

In this section we examine some of the problems facing an initiator implementation, and find very similar difficulties.

A very common form of human interface for file transfer is to allow the (human) user to request a file transfer between a local file and a remote file, to note the request on disk, and then (later) to progress the transfer on the user's behalf, sending failure or success messages to some mailbox.

Such an implementation faces the same problem as responder implementations in mapping document types to and from real files on the local system. The problem is, however, aggravated by the fact that *fragments* (single data units) of a remote file may be easily handled, whilst the whole file cannot be. The FTAM service and protocol is designed to identify support for complete document types, and some care is needed to provide initiator implementations with minimum constraints.

As with responders, an initiator must face up to some very difficult problems on character sets. Frequently the operating system manual defines a particular encoding as $, but actual devices on the system treat the encoding as £. *Neither* of these are the character graphic defined by ISO for the default use of this encoding in GraphicStrings, but both character graphics can be 'legally' transmitted by using appropriate escape sequences. To tell a lie (transmit without the escape sequence) or not to tell a lie (involving scanning text for the offending character) is an interesting decision! (Thus the problem, identified in the first chapter of this book, of $, #, £ confusion is unlikely to be totally eliminated by the coming of OSI!)

A fundamental choice for initiators is the sort of spooled approach defined above (good for simple transfers, but not for much else) compared with a highly interactive approach where the (human) user has available to him commands very closely related to FTAM service primitives, the primitives being issued as he types the command. (This is particularly suited to reading and changing attributes of a file.)

At first sight, this approach makes a 'full' initiator implementation much easier, but the problem of how to present to (or accept from) the user data values on F-DATA is a very difficult problem unless hex representation of the encoding defined for transfer is used, or fairly sophisticated configuration of abstract syntaxes and representations is employed.

Other major options relate to the provision of FTAM interfaces for terminal use or for programming language use, and decisions on whether to put recovery responsibility (reliable service) on the 'human' user – risking becoming a non-conforming implementation.

So far, we have not considered use of FTAM for real file *access* by editors and programming languages.

How should editors or Fortran I/O statements be extended to reflect access to the FTAM FADU structure and to general document types? At what level in an operating system should an 'escape' to FTAM for remote access be placed? Language subroutines? Language I/O subsystem? Operating system file access primitives? The answer will depend at least partly on the degree of match or mismatch between facilities for local file access and those provided by FTAM.

Thus, we see that even for initiator ends, the long-term pressure is to migrate the local system to match the FTAM virtual filestore.

Finally, some thought is needed on the problem of good user interfaces for access to the *full* power of OSI Application Layer protocols. Provision of simple-to-use interfaces for subsets of FTAM, JTM, and VT is fairly easy. Extending this for access to the full power of the protocols is much more difficult.

10.11 CONCLUSION

The FTAM Standard was close to stability when this chapter was written, and significant attention was beginning to be focused on implementation.

The Standard provides a highly flexible and powerful tool for remote access to information-handling devices, and is likely to be used in increasingly sophisticated ways into the 1990s.

The impact of FTAM on filing system design, the emergence of *de facto* standards for a variety of document types and file structures, and the flexibility of actual implementations will be areas to watch during this period.

CHAPTER 11

Job Transfer and Manipulation

This chapter describes the features of the Job Transfer and Manipulation (JTM) Specific Application Service Element (SASE). This is the technical term for the standardisation provided by ISO 8831 and ISO 8832, two important Layer-7 OSI Standards.

11.1 HISTORY AND OBJECTIVES

In the early days of computer use, all processing was performed as 'background' or 'batch' jobs, frequently submitted from decks of cards. As remote computer services developed, they provided their services by using physical mailing of card-decks and output to and from their customers.

There developed quite rapidly, however, equipment (originally special hardware, then later software emulators on general-purpose computers) to support *remote job entry* – card-reader and line-printer clusters remote from the host computer. Today there are few such pieces of equipment in use, at least for input, but the protocols they employed (U200, 7020, 2780) are still used to enable one computer to transfer background work to another, and to receive output.

Increasingly today we see work performed, not as background jobs, but in the foreground, supporting terminal users, with the terminals either local or remote. This mode of operation is suited to human–computer interaction, but is insufficiently standardised for the computer–computer communication which is now being required.

The OSI JTM protocol is designed to support computer–computer communications for the purpose of performing work remotely. In doing this, it recognises not only the old requirement for submitting 'batch' jobs, but equally the requirements of 'immediate' processing. Special cases of computer–computer communication with 'immediate' processing are described as 'transaction processing' and 'distributed databases', or more generally as 'distributed processing'.

JTM provides a type of communications support which covers some of the requirements for distributed processing. It is, however, optimised for the case of 'background' work where a (human) user specifies the total work to be performed (on a number of systems), and then leaves the computers to get on with it.

The JTM protocol can be usefully applied in support of traditional scientific computing where an initial transfer carries a piece of job control language (JCL), program, and data, with instructions (and perhaps passwords) for a later transfer to distribute resulting output.

It can also be applied usefully in support of a flexible manufacturing plant, where an initial transfer carries the specification of a part to be manufactured, with instructions (and perhaps authorisation codes) for a later transfer to advise of despatch and/or carrying an invoice.

The communication requirements, and communication-related data needs are the same in both cases.

It was to satisfy these needs that JTM was developed.

11.2   USE OF COMMITMENT, CONCURRENCY AND RECOVERY

The Commitment, Concurrency and Recovery (CCR) Standards have been described in section 9.3. They are essential for almost all activity which involves a number of computers and which wants to have reliable operation despite crashes.

Where CCR is not used, some degree of reliability is normally achieved by ad hoc CCR-like mechanisms (see, for example, the discussion of the FTAM reliable service in Chapter 10).

The JTM Standards operate at all times within a CCR atomic action, and rely on the CCR provisions for reliable operation, and for consistent processing across several systems.

11.3   JOB TRANSFER AND MANIPULATION OVERVIEW

The requirements of a JTM Standard can be developed in two separate directions.

Some early work in the UK concentrated on trying to standardise the 'job control language' or 'command language' by which local processing was controlled. This attempt was unsuccessful even for the very limited forms of processing covered by the term 'scientific background jobs'. The more general forms of processing addressed by ISO JTM would have been even more difficult to handle. ISO JTM work deliberately avoided this area. Standardisation here is a useful goal, but has limited concern with communication and OSI. Thus the ISO JTM specifies the means to move documents (collections of data) around. One document may be some form of local control language; another may be some form of program. It is of no concern to JTM.

The second direction, taken by ISO, is to focus on *information flows*, and

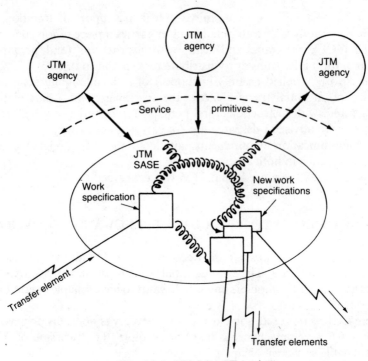

Fig. 11.1 JTM SASE actions.

the control of information flows. This leads, in due course, to a protocol which effectively *manages* transfers of documents.

Developing the theme of our examples, what do we see for the basic information flow requirements?

Clearly, the following steps can be taken in developing a JTM protocol:

Step 1 – recognise the need to form a document by concatenation of material from a number of different systems and pass it on to some target system for processing;

Step 2 – recognise that the target system will produce (over a period of time) a number of documents resulting from the processing;

Step 3 – recognise that information is needed by the target to control and permit the distribution of these documents;

Step 4 – recognise that in the general case these documents will be merged with documents from other sources, and perhaps need to be duplicated, to produce new documents for further processing at different targets;

Step 5 –   recognise that this chain (tree) of distribution, processing, distribution, processing ... will usually end with filing or printing 'ultimate' results, but that the number of steps is potentially unbounded, and could cycle in a repetitive fashion for some applications of JTM;

Step 6 –   recognise that the entire tree of activity *may* complete 'immediately', but will, in general, be subject to networks becoming available, resources becoming available (scheduling), and processing time (CPU cycles or manufacturing a part); in general, the activity will take some (potentially long) time to complete;

Step 7 –   recognise the need for reports (to some mailbox, file, or printer) to allow the (human) user to monitor progress of the activity if it is taking a long time;

Step 8 –   recognise the need to enquire about the activity (request the status of the work), and perhaps to modify some parts of the specification of later work, including killing it, and including correcting errors in filenames etc;

Step –   recognise the security problems of permitting authorised access to systems for an indefinite time in the future, and of authorising modifications of the work while preventing unauthorised changes;

Step 10 – recognise the problems of overload and multi-queue scheduling if many independent sites are all directing output to the same system;

Step 11 – recognise the problems of mobile systems, only occasionally interconnected, or (the same problem) of systems in grossly different time zones never switched on together.

The above chain of reasoning leads to the requirements that the JTM Standards attempt to satisfy. The result is a large and quite complex service specification, defined in ISO 8831. (Probably more complex and facility-rich than any of the other OSI protocols.)

11.4   JOB TRANSFER AND MANIPULATION SUBSETS

The need to recognise subsetting, or a 'Basic Class' appeared very early in the JTM work, but unlike other OSI standardisation activity, JTM attempted to define initially the *full* service, in order to ensure that the Basic Class provision was, indeed, a true and sensible subset of the general case.

The 'Basic Class' protocol (ISO 8832) is a kernel of essential functionality for simple JTM-type operations. The full protocol extends this functionality by the addition of one or more 'functional items'. Thus a deep tree of 'distribution, processing, distribution, processing …' is not in the Basic Class, but requires the extended functionality. Modification of uncompleted activity is not in the Basic Class. Concatenation and duplication of documents is not in the Basic Class. Reporting of events is limited in the Basic Class. Control of transfers to aid distributed scheduling is not in the Basic Class. Checkpointing of transfers to avoid complete retransmission on crashes is not provided in the Basic Class. Document collection from 'third-party' systems, either by use of FTAM or by direct JTM protocol, is not in the Basic Class (they have to be at the source or at the target).

Thus, Basic Class represents a reduced functionality which is both relatively simple and quick to implement, but is also adequate for many applications.

In terms of history, the functions in the full specification were largely proposed by Germany and the UK. The precise selection of features for Basic Class was predominantly the result of Japanese proposals. The USA and France made little contribution to this area of OSI, although they took a strong interest in CCR and more general aspects of distributed processing.

The JTM Standards recognise two independent dimensions of subsetting. The first was reflected in the above discussion of Basic Class functionality. The second recognises, in layman's terms, that an implementation may support any one or a combination of:

(a) specification of the work to be done (by human interaction at a terminal);

(b) receipt, storage and display of JTM reports for a human;

(c) specification of status or manipulation requests, and display of the results (to a human);

(d) a filestore to which JTM documents resulting from processing can be sent;

(e) a printer or plotter or other output device to which JTM documents resulting from processing can be sent;

(f) a processing system;

(g) the ability for user programs written in some language to generate JTM work, process JTM documents, or receive JTM results.

This additional dimension of subsetting, together with the Basic Class and extended functionality dimension provides a controlled but flexible approach to JTM implementation.

11.5 JOB TRANSFER AND MANIPULATION MODEL AND TERMINOLOGY

As with most Layer-7 OSI protocols, JTM defines a *model* of the functions of a real system. It expects an implementor to map parts of that model on to real devices – printers, tape-decks, filestores, plotters – but does not in any way constrain how he does that. Thus JTM does *not* talk about a 'printer', or a 'batch job system', or an 'order processing system', or a 'mailbox', or a 'human being'. Rather, it talks about a 'sink agency' (anything which takes documents) a 'source agency' (anything that provides documents), an 'execution agency' (anything that processes a document and later acts as a source of new documents), and an 'initiation agency' (anything that specifies JTM activity).

JTM defines *service primitives* (see Chapter 3) which interact with agencies as the result of JTM protocol exchanges. Later in this chapter we will examine the defined primitives, but broadly they are very simple. There is an outer shell of CCR-related primitives to bound and identify the atomic action, with a simple transfer of responsibility for a document as the inner core.

The most important concept in JTM is the so-called *work specification*. This is a data structure (collection of structured information) which completely specifies the patterns of document movement and processing (and subsequent movement) which the (human) user requires in this so-called *OSI job*. The work specification has a defined information content, but its detailed structural form (syntax) is only specified when the information is to be transferred. It can be held locally in whatever (distributed or centralised) fashion an implementor chooses.

JTM chooses to use the work specification concept not only for the basic document movements, but also for moving reports, manipulation (status or modification or kill) requests, and results of manipulation. Thus, once the service primitives and the work specification are understood, there is little else to understand (or to implement) for JTM.

Formally, JTM recognises the existence, at open systems distributed about the world, of:

(a) initiation agencies;

(b) sink agencies;

(c) source agencies;

(d) execution agencies;

(e) and JTM implementations (JTM SASEs) accessing and supporting these agencies.

An initiation agency interacts (via service primitives) with the JTM SASE to produce a work specification. This is processed by the JTM SASE, possibly resulting in interaction with source agencies to obtain documents for inclusion in the work specification. (Remember, work specifications are just collections of structured information – large documents are one possible component; real implementations will typically include documents by means of pointers, rather than physically copying data.)

The work specification may have an initial target (for processing a document) which is remote. The work specification (together with any documents it now contains) is therefore transferred (by the JTM SASE) to a JTM SASE on the remote system, using a defined syntactic form called a *transfer element* (defined using ASN. 1 – See Chapter 8).

The remote JTM SASE processes the work specification, resulting in further interactions with agencies, and often, eventually, in new transfer elements to other destinations. This is shown in Fig. 11.1.

11.6 JOB TRANSFER AND MANIPULATION WORK SPECIFICATIONS

The general form of a JTM work specification can only be described recursively. We need the concepts of:

(a) a subjob specification;

(b) a proforma;

(c) global parameters.

A subjob specification is information which determines where documents are to be collected from, and how they are to be concatenated, and to which agency or agencies on the (single) target system the resulting document(s) are to be passed. In full JTM, but not Basic Class, a work specification – transferred using transfer elements – can move round a number of *JTM relays*, each of which will examine the subjob specification and, if so required, interact with local agencies to collect documents and add them to the work specification before onward transmission towards the target. Alternatively, in full JTM but not in basic class, documents required by a work specification can be collected (by any JTM SASE implementing this feature) from remote systems by using the FTAM protocol. The work specification contains FTAM-specific fields to aid this activity.

A work specification contains some *global parameters* relating to the OSI job as a whole, and (most importantly) a top-level subjob specification which is processed by JTM SASEs, and also a top-level *Proforma list*. The proforma list contains one or more proformas (not affected by any initial processing) *each of which contains a subjob specification and a proforma list*.

When processing a subjob (including any related local processing of documents) is completed at a target, the proformas are 'spawned', i.e. they become independent work specifications in their own right, including their own copy of the global parameters. Proformas can also be turned into work specifications during this processing by 'demand spawning' from an execution agency (see Fig. 11.2).

An 'inheritance' mechanism allows a spawned proforma to have a copy of itself or one of its brothers inserted in it when it is spawned. This is a very powerful (and simple to implement) mechanism for recursive definition of

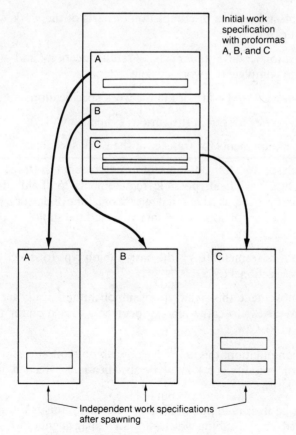

Initial work specification with proformas A, B, and C

Independent work specifications after spawning

Fig. 11.2 Spawning proformas.

JTM activity, but (at the time of writing) a real requirement for the facility has not yet been identified.

A work specification consists of global parameters (information affecting the entire activity), a subjob specification and a proforma list. Whenever a proforma is spawned, the global parameters are copied into the new work specification.

The global parameters, called by JTM the 'OSI job parameters', contain information about:

(a) the identity of the system and (human) user initiating the work;

(b) the time of initiation;

(c) identification of the work;

(d) tracing, for authentication security purposes, of the systems which have processed the work specification (or its ancestors);

(e) where to send reports of completion of parts of the work, warnings, diagnostics, etc.;

(f) authorisation codes, statements of authentication, and passwords needed to complete the work or deliver reports;

(g) who is allowed to modify or kill the work specification;

(h) accounts to be charged at the various systems involved;

(i) security requirements for transfer of the work specification.

The subjob specification contains information about the target (and any relays), an 'urgency' field, possible requirements to 'hold' the subjob pending some event (such as manual release or a time of day or a date–time or expiry of a timer for some fixed interval), and the subjob 'type'.

The 'type' is one of:

(a) document-movement: this is the basic subjob type for specifying 'real' work (see section 11.7);

(b) report movement: this is the type of subjob automatically generated by the JTM SASE to carry report documents to a monitor point (see section 11.8);

(c) work manipulation: this is the type of subjob requesting the target to report on, modify, or kill, work specifications which it holds (see section 11.9);

(d) transfer control manipulation (not basic class): this is the type of an advanced form of subjob which allows a management centre to control the transfer attempts of a JTM SASE (see section 11.10);

(e) report manipulation (not basic class): reports are normally sent to files, printers, or mailboxes; however, they can be sent for retention by a JTM SASE. The report manipulation subjob causes that SASE to display or delete the reports it has received (see section 11.11).

Depending on the type, the subjob specification then contains a detailed description of what is to be done, in the so-called 'JTM actions' field. This field is described further below.

11.6.1 Document movement work specification

A subjob specification with a type of document movement contains a JTM-action field which, at the end of processing by JTM SASEs (but prior to final processing by the target) contains *a list of* individual specifications, *each* of which:

(a) is associated (one-to-one) with a single document which is now present in the work specification;

(b) requires that document to be passed by the target JTM SASE to one or more sink or execution agencies which it can access locally, or remotely by use of FTAM.

(In basic class, there is a single document, a single specification of this form, and a single disposal agency.)

A subjob specification becomes active either as the top level of an initial work specification, or when a proforma is spawned. At this time it is either already in the above form, or contains specifications which are processed by JTM SASEs on the route to the target (including the initial one and the target) to get it into this form. (Again, the processing of subjob specifications by JTM SASEs in the basic class is limited and simple.)

There are two main initial forms which are processed into this final form.

For the first form, the single document (for disposal) associated with an individual specification starts off as a list of references to source agencies, each of which is to be accessed (locally or via FTAM) by a named JTM SASE in order to obtain a fragment of the final (single) document. The work specification is passed from JTM SASE to JTM SASE en route to its target JTM SASE, each one 'resolving document references' for which it is named.

For the second form, a *group* of individual specifications starts out as a single individual specification, with the 'document' identified as a group of documents at a single source or execution agency – typically all documents produced by processing, or all files in some directory. This form is processed (by the JTM SASE named as responsible) to produce a *list* of individual specifications, each requesting the target to pass one of the original group of documents to the list of sink or execution agencies. The names to be used for disposal of the documents are derived from their names in the source.

All this sounds complex, but an individual SASE has only three quite simple actions to perform (perhaps several times):

(a) resolve a simple reference to a source agency by getting a document and putting it in the work specification;

(b) get the names of a group of documents and replicate the appropriate piece of specification, including the document names in the new specifications;

(c) dispose of a document to a sink or execution agency, await a signal that any processing or disposal (printing) is complete, then spawn proformas.

There are details concerning *<agency access parameters>*, which allow, for example, control of deletion following receipt of a document (move or copy), and control of overwriting old files or appending to them. Further detail is beyond the scope of this text, and the reader is referred to the actual Standard.

11.6.2 Report movement specification

A subjob specification with a type of report movement is, by comparison, very simple, and undergoes no change from initial creation to final processing at the target.

Report work specifications cannot be created by spawning proformas, nor can they be created by human initiators. JTM identifies fifteen (for full JTM – four in Basic Class) events which can cause reports to be generated. (The first four listed below are those available in Basic Class.) Note that recognition of an event as a candidate for reporting does not mean that reports are always produced. The events are as follows:

Normal termination: a subjob specification has been completely processed by its target, and the work of agencies is complete (this report identifies any proformas which were spawned);

Abnormal termination: errors were detected in the specification, and the work had to be abandoned (see below for a discussion on error-handling in JTM);

Manipulation termination	the work was killed by a 'manipulation' (see later discussion);
User message:	a service primitive exists for execution agencies to pass messages (originating from processing a document) to the JTM SASE. These are not standardised, but are an 'event' which can be reported. The report naturally carries the message;
Creation:	the initial creation of the first work specification of an OSI job;
Transfer:	the transfer of responsibility for progressing a work specification to another JTM SASE;
Spawning:	a proforma has been spawned;
Agency acceptance:	a target JTM SASE has passed documents to agencies. The agency has accepted the document for disposal or processing, but the work of the agency is not yet complete (compare normal termination);
Modification:	the modification of a work specification by a 'manipulation';
Error diagnostic:	when a JTM SASE is attempting to transform a subjob specification by obtaining a document from a source agency, it may fail to do so (bad filename, inadequate authorisation, and so on). JTM supports several error actions (see later), but one of these is to pretend the action had succeeded, and allow the work to continue, but place in the work specification, instead of the requested document, an *error diagnostic*. This occurrence is an event which is a candidate for reporting. The report carries full details of the file being accessed and the error diagnostic;
No progress:	another possible error action (which is

applicable to obtaining documents, transferring work specifications, or disposing of documents) is to automatically place a 'hold' on the work specification to allow the (human) user to correct the error (create a file perhaps), and to release the 'hold' by a 'manipulation'. Placing the work specification on 'hold' is a 'No progress' event. Again, the report will carry details of why the work specification was placed on 'hold';

Accounting data:

this event is the availability of charging information which may require to be reported. It can occur many times in the life of an OSI job;

Not supported termination:

an implementation supports some combination of JTM functionality from basic class (all mandatory) up to full functionality. Thus, a user requesting JTM functions beyond basic class may be using a system where the functions are not supported. This gives rise to this event;

Violation attempt:

this event occurs if a 'manipulation' requests modification (or killing) of a work specification, but does not carry sufficient authorisation to do so. The report is sent to the monitor points of the work specification on which the attempt was made;

Warning report:

this is a catch-all. If warning messages are generated during a 'successful' activity, e.g. a sink agency folding or truncating long lines, these are an 'event' which can be reported.

Each work specification contains, as part of its global parameters (copied on spawning) a so-called *primary monitor point* and a list of *secondary monitor points*. These are sink agencies associated with a JTM SASE to which reports are to be sent. For each monitor point, there is a list of the events for which reporting is requested. The primary monitor point is inserted by the JTM SASE, not the initiation agency, and a 'manipulation' requires management authority (see later) to change it. Secondary monitors (and the selection of events to be reported to each one) are determined (and modified) by the user.

When an event occurs for which reporting to one or more monitor points has been selected, the JTM SASE automatically creates a new work specification (copying the global parameters) to carry the report to the monitor point. Further details on the form of reports is beyond the scope of this text.

It is important to note that the available reports are sufficient to enable a sophisticated monitor-point to keep detailed track of the status and location of the various parts of an OSI job. The detailed operation of such a monitor point is, however, not standardised.

11.6.3 Work manipulation specifications

These subjob specifications (which *can* appear in proformas, and hence be spawned) are also unchanged from creation to processing at the target.

They contain, in addition to a target and possible 'hold' elements, a *list* of operations. Each operation is one of the following:

Select: select one or more work specifications by giving values to match various fields. Almost all fields of a work specification can be used for selection, and wild-cards (matching any value in a field) are supported. The selected work specifications are subject to later operations up to the next 'select';

Kill: stops all processing associated with the work specification (by a service primitive interaction with associated agencies), and discards the work specification;

Stop: interacts with agencies (using a different service primitive) to cause them to cease processing, but any results so far are not discarded, and the JTM SASE continues to handle the work specification normally;

Modify: changes fields in the work specification (including values in proformas);

Display: a full or brief display of a work specification, and its current status (again involving service primitive interactions with associated agencies) is generated.

Any 'manipulation' requires the manipulating work specification to contain *authorisation elements* matching *permissions* in the work specification being manipulated.

A manipulation work specification causes a 'manipulation response' document to be generated by the JTM SASE. Following processing of the manipulation, proformas are spawned in the usual way, and collect this document for distribution exactly as for any normal document movement. Figure 11.3 shows some of the patterns of movement of work specifications which might be associated with a job. It includes 'report manipulation' which is covered in section 11.11.

11.6.4 Transfer control manipulation

There is one other important data structure used by JTM (in addition to work specifications). This is the *transfer control record*.

In normal operation, a JTM SASE makes purely local scheduling

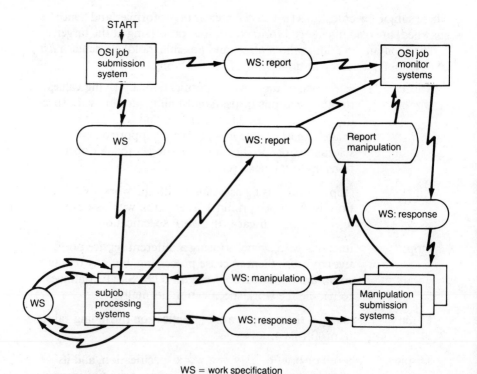

WS = work specification

Fig. 11.3 Work specification movement.

decisions about transferring work specifications. It may do one at a time. It may do twenty at a time. If it tries to send twenty to one site simultaneously it may work, it may always overload that site, or it may overload the site only if other sites are active.

A *JTM management centre* can attempt to control this situation. A common case would be where the management of *all* transfers *to* a particular site (site B, say) are controlled from site B itself by placing transfer control records at all sites A_1, A_2, A_3 ... that are known to have queues of material for B (as evidenced by transfer attempts, or by prior knowledge).

A transfer control record at site A, say, controls transfers to a single site B, and is independent of other transfer control records at A (if any) controlling transfers to different sites.

A transfer control record is simply a series of *selectors*, identical to those in the 'select' manipulation command. Each one is a kind of filter which allows through at the most one work specification at a time, and requires it to satisfy the selector. Thus, the number of simultaneous transfers from A to B never exceeds the number of selectors, and selectors can be set to allow, for example, only one big transfer at a time, but many small ones; or only one document movement but many reports; or only a work specification, or only work specifications which have been waiting more than a specified time; or various combinations of these basic approaches.

A transfer control manipulation subjob specification carries one of two operations. The first is sent by the management control centre to A (say) and is the 'set' operation. This sets, replaces or deletes the transfer control record at A for transfers to B (say). (A single transfer control manipulation can set (independent) transfer control records at A for a number of destinations B_1, B_2, B_3, ...)

The second operation is used by A, usually following recovery from some break in service. This is the 'check' operation, sent by A to the management control centre (the site that set one of its transfer control records), saying 'this is what I am using, is it still appropriate?'

An illustration of this activity is given in Fig. 11.4.

The general problem of distributed scheduling algorithms is a complex one. The JTM transfer control record provides a useful basic tool for enabling some remote control of transfer scheduling. Its importance will increase when/if user communities use JTM sufficiently for queues and overload situations to develop.

Experience with batch systems has shown that provision of basic tools for scheduling is a vital requirement in 'work-horse' (as opposed to 'toy') systems.

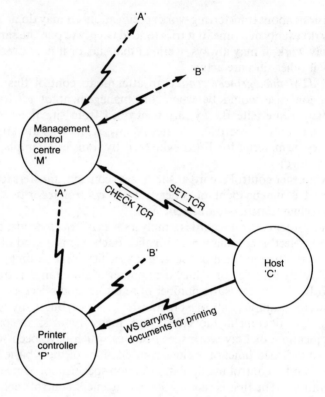

Fig. 11.4 Transfer control.

11.6.5 Report manipulation

So far, we have assumed that reports delivered to a JTM SASE were immediately passed to a sink agency (acting as a non-standardised monitor point) for filing, printing, or display via a mailbox.

JTM provides one further option. Reports can be sent to a JTM SASE for *retention*. The (human) user then, at some later time, initiates a work specification containing a report manipulation subjob. This requests either *display* of reports (returned, as usual, via a proforma), or *deletion* of reports. Reports for display or deletion can be selected by the name of the OSI job or subjob, the initiating site or user, the type of report (event), the type of the subjob being reported on, or any combination. As usual, authorisation mechanisms are fully developed, based on the 'permissions' carried in the report work specification delivering the report (copied from the OSI job being reported on).

11.7   ERROR-HANDLING

Mention has already been made of some of the features of JTM for error-handling. A variety of mechanisms are available.

For basic JTM activity (getting or disposing of documents, or transfer attempts) there are three options available on a per-subjob basis:

(a) embed diagnostics in place of the document and continue (obtaining documents only; this is the only option in basic class);

(b) hold for a specified time interval, and thereafter try again, terminating on failure;

(c) terminate with a diagnostic.

It should be noted that all CCR activity generates either 'retry-later' failures or 'no-retry' failures. 'Retry-later' failures (typically due to congestion) do not invoke any of the above mechanisms. They are simply tried again. Only after a long sequence of 'retry-later' failures are these treated as 'no-retry' failures which invoke the above mechanisms.

Another type of 'error' is when processing of documents by an execution agency takes an unexpected turn, resulting in no result documents, or different result documents. This affects the spawning that is appropriate.

Each proforma carries with it 'spawning control data' which control when it is spawned. The simplest case, 'completion', allows automatic spawning at the end of the subjob. There are three other options (but not in basic class):

Demand only:	the proforma is spawned only by a specific request from an execution agency (using a service primitive interaction) which is processing a document from the work specification;
Acceptance:	the proforma is spawned as soon as all documents have been passed for disposal or processing;
Conditional:	the proforma is spawned as for 'completion' but if (probably due to errors detected during the processing) it fails to pick-up at least one document at this site, it is immediately deleted as if it had never been spawned.

Finally, the use of 'user message' reports from execution agencies can provide for signalling exceptional events to monitor points.

11.8 AUTHORISATION

JTM recognises the need for a work specification to carry:

(a) a list of the 'usernames' or 'site names' (management) on whose behalf it is allowed to operate, and which authorise it to perform certain actions;

(b) some means of preventing masquerade in the use of such names, i.e. *authentication* information for the names;

(c) a list of 'usernames' or 'site names' which are given permission to affect it by a manipulation.

There are some 'permissions' which are automatic. Thus, a properly authenticated 'site name' (representing management at that site) always has the right to manipulate jobs originating from that site. Moreover, manipulation of report work specifications always requires this 'management' form of authorisation.

Authentication is most commonly conducted by use of a password (included with the name in the work specification). This mechanism is available in JTM, but it suffers from many problems, three of which are:

(a) the password is visible in transfers unless encryption is used;

(b) the password may be in the OSI job (and hence not be easily changeable by the user) for a long period of time;

(c) the authentication can only be performed by the site holding the password database, or by management protocols which do not yet exist.

JTM therefore introduces an additional mechanism which, in its simplest form, is *third-party authentication*. Instead of a password, the field contains a statement by one of the JTM SASEs which processed the work specification (or, commonly, one of its ancestors) saying 'this name has been authenticated'. *Provided* the work specification is known (to an adequate level of confidence) to have come from that JTM SASE without illegal changes (to an adequate level of confidence) by either the intermediate JTM SASEs or during transfer, *then* the statement can be accepted and the appropriate authorisation granted, e.g. to modify some work specification, or to delete a file.

This basic outline is supported by an 'audit trace' listing the JTM SASEs who have processed an OSI job, and concepts of 'authenticated', 'known' or 'unknown' for identifying the sender of a work specification which is being received across a network.

Further details are beyond the scope of this text.

11.9 JOB TRANSFER AND MANIPULATION SERVICE PRIMITIVES

The basic concepts and operation of JTM should now be well understood. It remains to describe the details of the service primitives by which JTM interacts with agencies.

JTM extends the service primitive notation described in Chapter 3 to recognise multiple 'users', not just two. Thus a 'request' primitive may produce 'indication' primitives in a whole series of application entities, each of which produce 'response' primitives which generate a single 'confirm'.

Every JTM interaction is within a set of primitives called J-BEGIN, J-READY, J-REFUSE, J-COMMIT, J-ROLLBACK and J-RESTART which are defined to be identical to corresponding CCR primitives (see section 9.3). Within each part of the atomic action, the JTM primitive interacts between a single JTM user (an agency) and a single JTM SASE. Interactions with other users may be going on in an interrelated way, but are described as separate primitives. CCR is used to ensure overall system integrity.

We will now ignore the CCR interactions and the global tree, and describe only the primitives occurring within the CCR exchange at one site. These are given below. Each branch of the atomic action includes only a single request/confirm or indication/response pair.

J-INITIATE request and confirm: the request passes a complete work specification from an initiation agency to the JTM SASE. The response returns a so-called 'OSI job local reference' generated by the JTM SASE (typically a time-stamp) which unambiguously identifies this OSI job within the JTM SASE for all time;

J-DISPOSE indication and response: the indication is issued by the JTM SASE and carries a document (or some error diagnostics), a document name, authorisation and accounting information, and information about overwriting, appending, etc. Another option provides the complete set of (FTAM-defined) information needed to dispose of the document using the FTAM protocol. This primitive is used for disposal to a sink *or* to an execution agency. In *both* cases, disposal/processing may complete 'immediately' (as part of the current atomic action, i.e. within the time specified by the atomic action timer) or the document may simply be secured. In the latter case there is an exchange of local identifiers (the only parameter on the response–refusal diagnostics are carried by the CCR-related primitives) which provide the association between the processing/disposal activity and the work specification. These are used in later service primitives;

J-GIVE indication and response: this is very similar to J-Dispose, except that the actual document is passed on the response, and the activity always completes or gives a retry-later or no-retry response. Thus, no exchange of identifiers is needed;

J-ENQUIRE indication and response: this primitive supports the conversion (by the JTM SASE) of a specification requiring all documents from a directory into specifications requesting individual documents. Information on naming, the types of document wanted, and available authorisations is passed on the indication, and the response returns a list of the names of available documents;

J-SPAWN Request: this is issued by an execution agency and identifies a

	processing activity (see J-DISPOSE for the exchange of identifiers) and carries a proforma name. It demands the spawning of the named proforma. A proforma can be spawned many times. It can also be spawned both on demand *and* on acceptance or completion;
J-MESSAGE request:	this is also issued by an execution agency, and carries a message arising from the processing;
J-END-SIGNAL request:	this is a signal (by an execution or sink agency) of completion of processing or disposal, where the J-DISPOSE gave only 'acceptance';
J-STATUS indication and response:	this indication (issued by the JTM SASE) identifies a processing or disposal activity, and the response carries a status message from the agency. This sequence, like the following ones, is invoked when processing a manipulation;
J-HOLD indication:	this informs the agency that a 'hold' has been placed on the work specification, and inhibits a J-END-SIGNAL. It is intended to suspend temporarily processing or disposal in the agency, but as agency behaviour is not standardised, this is not specified in detail;
J-RELEASE indication:	this informs the agency that a previous J-HOLD is nullified;
J-KILL indication:	this demands cessation of processing/disposal, and discard of any results;

J-STOP indication: this demands cessation of
                                      processing/disposal, but with
                                      retention of any results produced
                                      so far.

Some further clarification is needed of the *global* sequence of service primitives. When the first atomic action is initiated, the initiation agency specifies both an atomic action timer and a minimum *commitment level*.
    The commitment level is one of:

(a) Provider acceptance;

(b) Agency acceptance;

(c) Completion.

Within the time specified by the atomic action timer (this defines 'immediately'), the user will have received either:

(a) a No-retry refusal indicating an error, e.g. insufficient authorisation: or

(b) a Retry-later refusal indicating (usually) that congestion or lack of network access prevents the commitment level the user wants from being achieved; or

(c) an offer of commitment at or above the requested commitment level.

A commitment level of 'completion' (the highest) means that all necessary communications have happened; all remote service primitives have been issued; all agency disposals or processing has been done. This is a very common level to request for manipulations.
    A commitment level of 'agency acceptance' means that the initial subjob specification has reached its target, and that all documents have been passed to agencies, which have indicated acceptance of the documents. The remainder of the work is progressed as a series of atomic actions (initiated by the J-END-SIGNALS from agencies) with the lowest commitment level.
    A commitment level of 'provider acceptance' (the lowest) means simply that the JTM SASE has secured the work specification and will progress it later. Provided communication resources were available, the JTM SASE would normally (subject to local scheduling) have tried for a higher commitment level, but if this did not succeed within the required time, it would simply secure the work specification and offer 'provider acceptance'.

11.10   JOB TRANSFER AND MANIPULATION PROTOCOL

This section describes the protocol which provides the JTM service as described above. The basic class JTM protocol is extremely simple as far as transfer is concerned:

(a) open-up a connection using A-ASSOCIATE (see section 9.2);

(b) issue C-BEGIN to start an atomic action (see section 9.3);

The first presentation data value to be transmitted is a complete transfer element (a syntactically defined form of the work specification, specified using ASN.1), and each subsequent presentation data value is a complete document associated with the work specification.

(c) use CCR to complete the atomic action;

(d) begin a new atomic action for another transfer, or issue A-RELEASE.

In the full JTM protocol, checkpointing is possible, and a potential receiver can connect and request material, but the transfer is still basically simple.

The entire complexity of the JTM protocol resides in the procedures associated with processing a work specification, all of which have been outlined in the discussion of the service.

11.11   DOCUMENT TYPES

Little has been said about the sort of documents JTM can carry. Like FTAM (see Chapter 10), JTM is a basic carrier for any sort of document.

In order to have successful communication, however, somebody has to define the semantics of a document, e.g. lines of characters or binary octets, and a syntax for transferring them.

JTM includes in its Standard (and *recommends* all implementations to support) the definition of three document types. The first is 'simple ISO text document' – an indefinite number of lines of indefinite length from *any* character set registered with ISO (this includes Japanese, Chinese, Urdu, etc.). The second is 'simple ISO print document' – including page-throw, double-line-space and overprint. The third is 'simple ISO binary document' – a single unlimited string of bits (not restricted to multiples of 5, 6, 7, or 8).

There is a full definition of named abstract and transfer syntaxes for these document types, making them immediately usable. Other document types, e.g. carrying graphical or facsimile information, need defining and naming and recommending (at least) to JTM implementors. Such additional types are likely to be added slowly.

11.12   CONCLUSION

The full JTM service and basic class protocol were stable when this text was written, but the full protocol has yet to be developed, and few details have been given in this chapter.

The protocol is strongly based on a protocol in use in the British academic community for scientific computing, so some relevant implementation experience is available. Applications in the commercial/business sector have been studied, but not yet implemented.

Provided the effort is made to produce full and flexible implementations, JTM should provide a very powerful tool to support a significant number of distributed processing applications.

CHAPTER 12

Virtual Terminal Service and Protocol

12.1 INTRODUCTION

This chapter covers the ISO OSI Standards under development which are intended for the control of human interactions with other humans or computers via terminals. Although human interaction has been the prime motivation for the development of these Standards, the Standards are not restricted to this purpose and may be used for direct computer–computer communication if desired.

The OSI Virtual Terminal Standards are based on the idea that the facilities offered by various real terminals can be described in a common abstract manner. It should then be possible to map each abstract facility on to a real facility of each real terminal.

This is in contrast to the approach taken by CCITT for their Recommendations X.3, X.28 and X.29 (often referred to as 'XXX' or 'Triple X' for short). The XXX Recommendations were developed to define how to support a simple terminal over a Packet Network offering X.25 access protocols. The simple terminal does not have the capability to support the full X.25 protocol and so it is connected to a packet assembler/disassembler (PAD) which is connected to the X.25 network as shown in Fig. 12.1.

The PAD communicates with the terminal according to Recommendation X.28 which specifies how the terminal user can 'talk' to the PAD to set up or release connections and/or change PAD parameters. The remote computer 'talks' to the PAD across the X.25 network according to Recommendation X.29 which also specifies how the computer can establish or release connections and change the PAD parameters. X.28 and X.29 also distinguish data which must be passed transparently between computer and

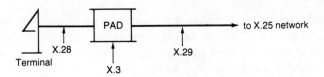

Fig. 12.1 CCITT XXX terminal standards.

terminal from the data which is intended to form the communications with the PAD itself.

X.3 defines a set of parameters which can configure the PAD to suit the terminal and the style of communication required by the computer.

ISO has not followed the XXX approach because:

(a) XXX is mounted directly on to X.25, a subnetwork access protocol, whereas terminal protocols are seen in OSI as being part of the Application Layer and therefore requiring intervening Layers 4 to 6 between network and terminal protocols;

(b) the Application process in the computer must in general be aware of the type of terminal connected to the PAD when using XXX;

(c) an ever-increasing number of X.3 parameters is needed to cater for existing and new terminal types.

It is believed that the OSI Virtual Terminal approach will isolate Application processes to a much greater extent from the variations in terminals and will need fewer and more rationalised sets of parameters than the XXX approach.

The rich variety of terminals and terminal facilities has made standardisation difficult, even following the Virtual Terminal approach. Terminals today come in black (or blue or green!) and white, colour, text-only, text with block graphics, bit-mapped graphics, raster scan graphics and many more features. Terminal technology is also developing rapidly and within the normal standards-making timeframe many new developments can be expected. In order to make progress within ISO, it was found necessary to define and concentrate on a basic set of terminal facilities and progress those independently of any more advanced facilities. Thus, after an initial attempt to try to define a generic virtual terminal model capable of adaptation to the majority of terminal features, a 'VT Reassessment Panel' was established and given a brief to define priorities and a programme of work for the VT working group.

At the time of writing, the VT group have defined a 'Basic Class' service and protocol (ISO 9040 and ISO 9041 respectively), together with an 'Extended Facility Set' (DAD1 to ISO 9040 and DAD1 to ISO 9041). It was originally foreseen that there would also be a 'Forms Mode' Standard but it is no longer clear whether there will be Standards in this area and if so when.

Recently, the use of computers for automation of personal and office activities has given rise to new requirements from computers. One important change has been the need for terminals to be able to handle several concurrent activities, e.g. so that the user can scan one document whilst writing another and then suspend both temporarily and look into one's diary or schedule. A number of terminals and personal computers have started to offer 'window' facilities whereby the terminal display is subdivided into discrete areas, or windows, and each is allocated to a different application.

ISO have just started to consider a new work item called 'Terminal Management' to standardise these facilities so that software packages can use these facilities without being too closely tied to any piece of hardware. This work item has not as yet been approved. Candidates for possible future Standards are forms mode, graphics, document and local editing enhancements. However, the facilities offered by the Extended Facilities addendum to the Basic Class cover much that would be thought to belong in a Forms Mode since the Extended Facilities allow for local entry and validation of data into fields to be performed by the terminal. It remains to be seen what if anything will happen in ISO with regard to Forms Mode.

The remainder of this chapter will be concerned solely with the Basic Class and its extended facilities as defined in ISO 9040 and ISO 9041 and their addenda.

12.2   THE VIRTUAL TERMINAL MODEL

It has already been observed that there is a growing diversity of terminals and terminal facilities and that the present work of ISO is directed towards standardisation of a well-defined basic subset, the Basic Class. The Basic Class is restricted to:

(a) character and character box graphics;

(b) simple 1, 2 or 3 dimensional arrays of characters;

(c) emphasis and colour of characters and background.

There is no substructuring of the display such as occurs in some page-mode terminals where groups of character positions are structured into fields with associated properties such as 'numeric', 'date', etc.

12.2.1 The Virtual Terminal Model elements

When a human interacts with a computer via a terminal, he is normally presented with some form of display and an input device, e.g. keyboard, bar code reader, 'joystick' or 'mouse'. As he uses the input device, the contents of the display are affected in some way and when the computer responds the contents of the display are also affected. Thus, the human and computer are both making changes to the display. Both have to be aware of the contents of the display so that their operations do not interfere with each other.

The VT Standards are therefore modelled as two entities altering a shared data area with each entity being made aware (eventually) of the alterations made by the other. The shared data area is called the 'conceptual communications area' (CCA). In practice, each entity will keep its own local version of the CCA and update it when notified of changes made by the other entity.

Because many real terminals have local memory which is greater than the capacity of the physical display and allow the human user to scroll or page through the terminal memory, the VT Standards decouple the conceptual display, i.e. terminal memory, from the real display, i.e. what is visible. The conceptual display is called a 'display object'. The mapping of the display object on to a real display is conceptually performed by a 'device object' (Fig. 12.2).

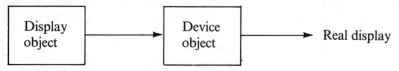

Fig. 12.2 Mapping display objects to real displays.

The display is defined to be a one-, two- or three-dimensional array. The one-dimensional case corresponds to a single line display, the two-dimensional case to a multi-line screen display, and the three-dimensional case to a multi-page display. The VT service describes each element of the display object array as having a primary attribute and a number of secondary attributes. The primary attribute is a character selected from a repertoire 'character set'. The VT Standards do not define the character repertoires. Implementations are expected to make use of existing Standards for character sets and graphics characters such as ISO 646 and ISO 2022. Secondary attributes include foreground colour, background colour, emphasis and font.

Many terminals also have alarms in the form of lamps, buzzers or other devices. These are generally used for the control of the terminal and its operations. The VT Standards therefore describe the conceptual communications area shared between the two entities as containing 'control objects' which can be defined by negotiation at the start of a communication but to which the VT service and protocol attach no semantics. Control objects are also seen as being mapped on to real devices by device objects (Fig. 12.3).

A control object contains an information field which is either a character string, a Boolean string, a scaler value or a bit-string. Updating a control object can optionally be controlled by an access control (see below).

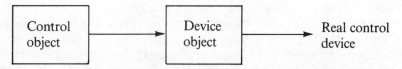

Fig. 12.3 Mapping control objects to real devices.

Each control object which is subject to access control may also optionally have a 'trigger' mechanism associated with it. The action of the trigger is to ensure delivery of all updates associated with the access right controlling the control object and, depending on the type of access control, to cause passing of access rights to the peer entity.

The conceptual communications area contains two other subdivisions:

(a) the access control store (ACS) which holds 'tokens' conferring access rights to each modifiable object in the conceptual communications area. Use of this token mechanism ensures that the two VT service users cannot both try to update objects at the same time;

(b) a data structure definition (DSD) which contains the definitions of the display, device and control objects in the conceptual communications area together with the values of various service parameters negotiated for the VT connection.

12.2.2 Variations of styles of use

We have already observed that there is a very wide variety of terminal types and characteristics which must be catered for by the Virtual Terminal Standards. There are also various different styles of operation possible and the Standards have to cater for these too.

The major variation is between a style of working which is two-way alternate, i.e. each VT user takes it in turn to access the conceptual communications area, and that which is two-way simultaneous, i.e. each VT user can access and modify the contents of the conceptual communications area concurrently. The VT service describes two modes of operation:

(a) the A-mode (*a*synchronous) corresponds to the two-way alternate style of working;

(b) the S-mode (*s*ynchronous) corresponds to the two-way simultaneous style of working.

A-mode is much simpler to operate and control and is thus to be preferred in cases where the restriction to two-way alternate operation is acceptable for the Application. Most normal enquiry/response applications are naturally two-way alternate, for example.

There are cases where two-way simultaneous working is needed. An example of this would be the control terminal for real-time management of a complex system such as a network or chemical processing plant. For these applications the terminal must be capable of being updated rapidly with status changes even if the operator is typing in a command.

In fact, the two modes had their origins in two of the more common types of terminal, the DEC VT100 (A-mode) and the IBM 3270 (S-mode) and represent the typical working styles of the mini computers and mainframe computers respectively.

There are two major effects on the operation of the VT service which follow from the choice of mode:

(a) in S-mode it is necessary to distinguish two separate display objects in the conceptual communications area, one associated with each VT user. In general, these display objects can be thought of as the keyboard and the display of a terminal. Other realisations are possible of course, e.g. there may be a mouse, joystick or other form of user input device rather than a keyboard or in the case where the VT protocol is being used for computer–computer communications there will be no physical device at all corresponding to keyboard or display;

(b) the style of access control is different in each mode. In A-mode the access control token passes alternately to each user, whereas in S-mode there are (conceptually) two tokens, one associated with each display object and each one permanently assigned to a VT service user. In A-mode, the single access control token is called Write Access Variable (WAVAR) and in S-mode the two access control tokens are called Write Access Connection Initiator (WACI) and Write Access Connection Acceptor (WACA). These names are meant to signify that possession of the token varies (WAVAR) or that possession of the token is permanently with the VT connection initiator (WACI) or permanently with the VT connection acceptor (WACA).

Because in the VT protocol the A-mode uses the two-way simultaneous services of the Session Layer and the S-mode uses the two-way alternate services and because the services of the Session Layer cannot be renegotiated on a connection, it is not possible to swap between A-mode and S-mode on one VT connection.

In addition to the two modes of operation the VT service also supports three modes of 'delivery control'. Delivery control allows a VT user to control delivery of data to the peer VT user and thereby co-ordinate multiple actions. An example of this might be an application where a single terminal key depression simultaneously causes a display of the key pressed and redefines 'prompts' on the display and the status of an indicator lamp. These actions can all be made subject to delivery control so that the peer VT user is presented with them all simultaneously. Another example is the use of special 'function keys' often found on terminals and which can be set-up to perform multiple actions. Sometimes it is desirable that all the functions of a key are presented to the peer VT user simultaneously.

The three styles of delivery control allowed are:

(a) **No delivery control.** In this case data is made available to the peer VT user at the convenience of the protocol implementation;

(b) **Simple delivery control.** In this case the VT user can issue a service request (the VT-Deliver service element) which causes all undelivered VT data to be delivered. There can be cases where the VT user will wish to ensure that a related sequence of updates are all delivered together. The simple delivery control service will not be suitable for this purpose because the VT service provider is free to deliver updates to VT users at any convenient time prior to the VT-Deliver;

(c) **Quarantine delivery control.** In this case the VT-Deliver service elements bound groups of SDUs such that delivery to the peer VT user must not occur before the following VT-Deliver. When using this form of delivery control, a process known as 'net-effecting' can occur. This is a process whereby only the final result of the execution of a sequence of SDUs (between VT-Deliver service elements) is present in the display object of the conceptual communications area when delivery is made to the peer entity. Net-effecting is intended to be an optimisation feature useful when a sophisticated terminal is being used over a relatively expensive communications line. It will be especially useful when terminal functions such as local editing are incorporated into the VT Standards where only the final effect of corrected entries need be passed to the remote peer entity.

There are two further optional styles of working, namely echo control and termination events. Both are included in the VT service because it is recognised that negotiation and agreement on these options must be supported by the protocol. However, both options relate to VT *user* actions and since the VT Standards cannot mandate requirements for the users it is not a protocol violation if the VT users do not, in fact, behave as agreed.

Echo control is specific to the A-mode and is concerned with the control of how characters typed on a keyboard will cause updates to a display. Because there are two independent display objects in the A-mode, actions on one object do not automatically cause corresponding actions on the other. In real terminals, characters typed on the keyboard may be displayed on the screen locally by the terminal or may be 'echoed back' to the screen by the computer. The former option is less flexible but is often chosen when the communications line is half-duplex and echoing back would therefore not be practical. The latter option is used where the communications line is full-duplex and where greater control over the screen is required, e.g. in order to suppress the display of the terminal user's password but display all other characters typed by the terminal user.

Termination events define, for a display object (usually the 'keyboard'), conditions under which the VT user is expected to invoke the VT-Deliver service element. It is expected to be used in conjunction with, for example, special function keys or to generate actions on timeouts such as enforced logout.

12.2.3 Terminal variations: Virtual Terminal Parameters

The possible variations in terminal characteristics within the Basic Class are catered for by a number of parameters whose values are decided by negotiation. The principles of this negotiation are covered later.

The parameters are organised into a hierarchy represented in the VT service Standard as a directed graph. This hierarchical organisation has been chosen because the choice of value for a parameter at one level in the hierarchy affects the number and types of parameters required at the next level down. For example, the choice between WAVAR and WACA/WACI styles of access (which are the means whereby S-mode and A-mode are selected respectively) will dictate whether there will be one or two display objects whose parameters must be defined.

Figure 12.4 shows the hierarchical relationship between parameters.

At the highest level is 'class'. At present the only valid value for this is 'basic'.

The parameters define:

(a) access rights (WAVAR or WACI/WACA) and thereby select S-mode or A-mode;

(b) the VT service subset (see section 12.3.2);

(c) the display object(s) characteristics (dimensions, array element attributes);

(d) the control objects (access control, trigger, information content);

(e) the device objects and their descriptions (access control, associated control objects, assignment of device characteristics to associated display/control object characteristics, termination events);

(f) type of delivery control.

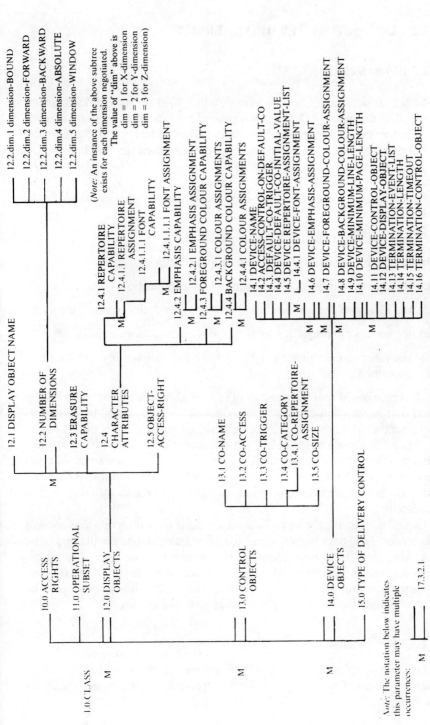

Fig. 12.4 VT parameters; directed graph.

12.3   THE VIRTUAL TERMINAL SERVICE

12.3.1 Main features

The VT service, Basic Class, is defined in ISO 9040 and an extended facility set in the addendum AD 1. It is not clear at the time of writing whether the addendum can or should become combined with the ISO text because the extended facilities require *restrictions* of the basic facilities. In other words, some facilities permitted by ISO 9040 would no longer be permitted by a combined text.

The VT service is a connection-mode service which provides the VT user with:

(a) the means to establish a VT connection;

(b) the means to establish agreed parameter values;

(c) the means to modify the conceptual communications area;

(d) the means to control notification to the peer entity of modifications;

(e) the means to control the integrity of the connection;

(f) the means to terminate the VT connection, unilaterally or by mutual agreement.

It is recognised that the number of (combinations of) parameters is large and that negotiation of parameter values individually could be very time- and bandwidth consuming. Thus the concept of 'profiles' is introduced.

A profile is a complete set of self-consistent parameters. It is expected that profiles will be developed to describe the most common terminal types and these profiles will be registered, i.e. will be given a unique name and have the parameter values recorded, along with the name, by a registration authority.

It will be possible for a profile not to include the values of some or all of the lowest-level parameters. These undefined parameters are defined to be 'profile parameters'. For example, it is common for a number of terminals to offer the same facilities and modes of operation but to vary in the number of lines or columns on the display. Thus, the display dimensions would be left as parameters in the profile for that group of terminals.

12.3.2 Service subsets

There are three major subsets of VT service which are distinguished by differences in the ways in which service and VT parameters can be negotiated.

VT-A, the kernel subset, will only allow parameter values to be established whilst the VT connection is being established. The VT parameter values are therefore fixed for the duration of the VT connection.

VT-B, the simple negotiation subset, additionally allows a simple 'switch profile facility' which can be used to alter the parameter values during the lifetime of a VT connection. The switch profile mechanism involves a single confirmed service element which allows a VT user to propose the use of a registered service profile and for the VT service provider and the peer VT service user to modify those parameters which are defined as profile parameters.

VT-C, the multiple negotiation subset, provides the switch profile facility and in addition provides a much more complex negotiation of parameters which allows for the modification of all parameters of a profile. In this case both the proposer of parameter values and the receiver of proposed parameter values are given the opportunity to accept or reject the parameter values.

12.3.3 Service elements

The VT service is comprised of a number of service elements each of which may be:

(a) confirmed (requiring the receiving VT user to respond to the indication with a 'response' primitive) or unconfirmed;

(b) sequenced (in which case all service primitives issued by the sender before this one are delivered as indications before this one is delivered) or unsequenced (may be delivered as an indication before some previous service primitives are delivered);

(c) destructive (in which case previously issued service primitives may be destroyed and never delivered by the service provider) or non-destructive.

12.3.3.1 Virtual Terminal connection establishment facilities

A VT user initiates a VT connection by issuing a VT-Connect request primitive. This is a confirmed service so the addressed VT user must respond.

The VT-Connect request contains a number of parameters:

(a) initiator's address;

(b) target address, i.e. the desired destination address;

(c) VT class (only 'basic' allowed);

(d) VT-facility-set (VT-A, VT-B or VT-C);

(e) VT-class-subset (only 'initial' allowed in ISO 9041);

(f) VT-access-rights ('WAVAR' or 'WACI-and-WACA');

(g) VT-WAVAR-initial-owner (only if access right is WAVAR);

(h) VT-profile-name (optional name of registered profile);

(i) VT-profile-param-offer-list (suggested values or ranges of values of profile parameters).

The VT-Connect is notified to the target VT user as a VT-Connect indication. At that time the value of the VT-class-subset and/or the VT-profile-param-offer-list parameters may have been altered by the VT service provider. Class subsets are defined to form an ordered set and the VT service provider and the VT connection acceptor are constrained to select a new class subset which occurs earlier in the ordered subset list. Modifications to the VT-profile-param-offer-list are constrained to those which reduce the choice from that offered and cannot introduce new choices, i.e. can only reduce a range or remove items from an explicit list of options offered for a profile parameter.

The accepting VT user (which need not be the VT user addressed by the target VT user address parameter) must complete the specification of the parameters by choosing values for each profile parameter where a range or list of options was given by the initiator. A VT-Connect response is returned which indicates success or failure of the connection establishment and contains the following parameters:

(a) responding address (the address of the responding VT user);

(b) VT-class-subset (may be a lower subset than that contained in the VT-Connect indication);

(c) VT-WAVAR-initial owner (may be different from that proposed in the VT-Connect indication);

(d) VT-profile-parameter-initial-value-list (lists explicit values for all profile parameters not (fully) defined in the indication;

(e) VT-result (success-with-warning or failure);

(f) VT-user-failure-reason (if the VT user rejected the connection);

(g) VT-provider-failure-reason (if the VT service provider rejected the connection).

The VT-result 'success-with-warning' occurs if the VT service provider or the responding VT user cannot find acceptable values for one or more of the profile parameters within the range or list offered in the VT-Connect request or indication. The initiating VT user can examine the response and determine from the still-undefined parameters which of them were unacceptable. In the case of VT-B or VT-C service subsets, it will then be possible to continue with the VT connection by using the switch context or multiple negotiation facility.

If the VT-result parameter indicates failure then a failure reason is given by the VT service provider or the responding VT user which rejected the connection.

12.3.3.2 Virtual Terminal termination facilities

Three service elements are provided for closing-down a VT connection which vary in their immediacy and destructiveness:

(a) **VT-Release** provides an orderly termination, requested by one VT user and agreed to by the other VT user, which does not cause loss of any data. The issuing VT user must own the appropriate token and when issued causes a forced delivery of any undelivered data to the recipient of the VT-Release indication. The recipient must return a VT-Release response which contains a VT-result parameter indicating success or failure. If the result is failure the VT connection remains open. If the result is success, then the VT-Release response causes forced delivery of data to the recipient of the VT-Release confirm and when the confirm is received the VT connection is closed. It may happen that the peer entity issues a VT-Release request at about the same time (specifically, between the issues of the VT-Release request and before receipt of the corresponding indication), causing a 'collision'. In this case, which can only occur in A-mode, the acceptor of the original connection establishment will receive a VT-Release confirmation from the VT service provider indicating failure due to collision and followed by the VT-Release indication corresponding to the request from the connection initiator. No service primitives may be issued by the sender of the VT-Release request before receipt of the corresponding confirm. The peer VT user cannot issue any service primitives between receipt of the VT-Release indication and issuance of the corresponding response.

(b) **VT-U-Abort** immediately terminates the VT connection, with possible loss of data, upon request from one VT user. VT-U-Abort may be requested at any time by either user. No further service primitives can be issued subsequently. It can be both unsequenced and destructive and is not confirmed by the recipient. In the case of a collision of two VT-U-Aborts, the corresponding indication primitives will not be delivered.

(c) **VT-P-Abort** immediately terminates the VT connection, with possible loss of data, upon the initiation of the VT service provider. Both VT users receive a VT-P-Abort indication with a VT-reason parameter to inform them that the VT connection is closed. This service primitive is used when an irrecoverable exception condition has occurred, either in the VT service provider itself or in the lower layer services. The VT service provider will release the lower layer connection to release lower layer resources as appropriate.

12.3.3.3 Virtual Terminal negotiation facilities

The set of VT parameters which define a given profile are called a VT environment (VTE). When such a set is complete and self-consistent it is called a full VTE. In the Basic Class, there can only be one full VTE established at one time. It is anticipated that extensions will permit more than one full VTE to be agreed and current at one time. The VT user will then be able to swap rapidly from one to another easily and rapidly. Negotiation facilities in the Basic Class are thus restricted to the modification of that single VTE (originally defined during VT establishment).

Two methods for changing the VTE are provided but their availability to the VT users depends on the service subset negotiated during VT connection establishment (see above).

The VT-Switch-Profile service element, available in subsets VT-B and VT-C, allows a VT user to initiate selection of a new VTE with a single confirmed service element. The VT user issues a VT-Switch-Profile request containing the following parameters:

(a) VT-profile-name;

(b) VT-profile-param-offer-list

The recipient of the corresponding VT-Switch-Profile indication issues a VT-Switch-Profile response containing the following parameters:

(a) VT-param-value-list;

(b) VT-result;

(c) VT-user-failure-reason;

(d) VT-provider-failure-reason.

The negotiation proceeds exactly as for the negotiation embedded within the VT-Connect service element with one exception, namely that the result 'success-with-warning' is not available. If the result cannot be completely successful then the result must be 'failure'.

The initiator cannot issue any service primitives between the VT-Switch-Profile request and receipt of the corresponding confirm. The peer VT user cannot issue any service primitives between receipt of the VT-Switch-Profile indication and issuance of the corresponding response.

The multiple interaction negotiation (MIN) service element is considerably more complex than the switch profile (SP) service element. The following service elements are involved:

(a) VT-Start-Neg: used to enter the negotiation phase;

(b) VT-Neg-Invite: used to solicit a proposal for a parameter value from the peer VT user;

(c) VT-Neg-Offer: used to propose a parameter value, either unsolicited or in response to a VT-Neg-Invite;

(d) VT-Neg-Accept: used to accept an offered parameter value;

(e) VT-Neg-Reject: used to reject an offered parameter value;

(f) VT-End-Neg: used to terminate the negotiation phase.

Each of the above service elements, excepting VT-Start-Neg and VT-End-Neg, may contain values or proposals for values of more than one parameter. There are constraints that apply to the sequences with respect to the negotiation of each parameter and the sequence for each parameter can be interleaved with those for other parameters and be combined into occurrences of the service elements listed above.

The constraints for negotiation of a parameter value which apply once the negotiation phase has been initiated by the confirmed VT-Neg-Start service element are as follows:

(a) in the S-mode, negotiation service primitives can only be issued by the holder of the WAVAR token;

(b) in the A-mode, the VT user which successfully initiated the negotiation must start all negotiation sequences for each parameter and must issue the VT-End-Neg request. The negotiation sequence for a parameter can be initiated by making a proposal for the parameter using a VT-Neg-Offer request primitive or it can be initiated by issuing a VT-Neg-Invite. The negotiation sequence for a parameter ends with either a VT-Neg-Accept or a VT-Neg-Reject. Figure 12.5 shows the possible sequences on a VT connection for the negotiation of a single parameter value.

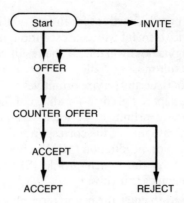

Fig. 12.5 Parameter negotiation sequences.

The final ACCEPT represents a successful outcome for the parameter in which a parameter value has been agreed by both VT users. The final REJECT represents an unsuccessful outcome for the parameter in which the VT users have failed to agree to a parameter value.

The negotiation phase is terminated with the VT-End-Neg service element. The VT-End-Neg primitives contain parameters which allow the users to determine whether or not to switch to use of the newly negotiated VTE and whether or not to retain the new VTE. A VT-Result parameter is also provided to indicate whether the VT provider and peer VT user were able to leave successfully the negotiation phase (VT-result = 'success') or not (VT-result = 'failure').

12.3.3.4 Data transfer facility

The VT-Data service element provides the means for a VT user to update the contents of the conceptual communications are a (CCA). One VT-Data primitive can contain updates to one or more objects in the CCA but all objects updated by one VT-Data must be of the same type, i.e. 'controlled' or 'uncontrolled'.

The parameters of a VT-Data primitive are:

(a) a (list of) VT object update(s);

(b) optionally a VT-echo-now parameter. If present it indicates that if there are updates waiting to be echoed, a suitable place for the receiving VT user to perform the echos will be after processing of the updates in this VT-Data.

Each VT-object-update consists of a VT object name and the data with which it is to be updated. In the case where the object is a display object the update data can include any combination of primary attribute, secondary attribute, pointer update and erase operation.

The pointer identifies an element of the display object array and various types of update are possible. The basic pointer updates are:

(a) 'pointer absolute' where an explicit value for each array dimension is given;

(b) 'pointer relative' where a signed *change* to the previous pointer position is given;

(c) 'home' which sets X, Y and Z values to 1.

In the above addressing operations, not all of X, Y and Z values need be specified and if they are not specified then they will be left unchanged. The following addressing operations can only be used as specified:

(a) 'new line', i.e. next X-array (X:=X min, Y:=Y+1);

(b) 'previous line', i.e. previous X-array (X:=X min. Y:=Y-1);

(c) 'next page', i.e. next Y-array (X:=X min, Y:=Y min, Z:=Z+1);

(d) 'previous page', i.e. previous Y-array (X:=X min, Y:=Y min, Z:=Z-1).

Where X min and Y min are the lower bounds set for X and Y by the defined window in force (see below).

The pointer position only affects the position where following updates will be applied. It does not necessarily correspond to the cursor position on the real display device. The concept of a 'window' is applied to the display object because terminals have limited local storage. When a window size (t) has been defined for an array dimension, only array elements between ('max'-t+1) and ('max') may be addressed by a pointer update operation (where 'max' corresponds to the highest value of that dimension that has been used in any update operation). The window sizes for each dimension of the display object array are specified independently as VT parameters. Note that when the primary attributes of a display object element is updated there is an implicit pointer update to the next x-dimension position (equivalent to a pointer relative operation of (X:=X+1, Y:=Y, Z:=Z)).

Erase operations cancel the assignment of primary attributes to one or more display object elements. There are three variations of the erase operation:

(a) erase line;

(b) erase page;

(c) erase book.

Variations of each of these are possible allowing for erase-all, erase-back (from the start of the current line, page or book to the current display pointer) and erase-forward (from the current display pointer to the end of the line, page or book).

Attribute updates alter a named attribute with one of the negotiated attribute values from the negotiated reportoire and operate over one of the following ranges:

(a) global, i.e. the new attribute value is applied to every element in the display object;

(b) forward, i.e. the new attribute value is applied from the current display pointer forwards to a defined end-position;

(c) backward, i.e. the new attribute value is applied from the current display position backwards to a defined end of range;

(d) explicit range, i.e. the new attribute value is applied from a defined start position to a define end-position;

(e) modal, i.e. the new attribute value is added to a set of such attributes which are then applied to each element of the display object which is altered by a text operation.

For the erase and attribute operations, normal wrap-around is assumed, i.e. when the end of a line (X-array) is reached the 'next' element is the start of the next line and when the end of a page is reached the 'next' element is the start of the next page.

When the VT-Data updates a control object which has an associated trigger, the update will imply a forced delivery of any outstanding updates of objects associated with the control object (in the case where delivery control is in force).

12.3.3.5 Delivery control facility

There are two service elements involved in delivery control:

(a) VT-Deliver: this primitive forces the VT service provider to deliver any outstanding updates. An optional parameter can request an acknowledgement of delivery. The VT-Deliver also provides a mark in the stream of VT-Data primitives since it is a sequenced primitive. A VT user which issues a VT-Deliver request with an acknowledge request parameter is forbidden to issue any further VT-Data primitives until the corresponding VT-Ack-Receipt indication has been received.

(b) VT-Ack-Receipt: acknowledges the receipt of a VT-Deliver primitive containing an acknowledgement request parameter. It does not force delivery of updates and, although it is sequenced with respect to VT-Switch-Profile and VT-Release primitives, it need not be sequenced with respect to VT-Data primitives.

12.3.3.6 Token management facility

For S-mode operations, the access control token can be passed from the current owner to its peer VT user by means of the VT-Give-Tokens primitive. In Basic Class there is only one token so there are no parameters for this primitive. The VT service provider records the change in ownership of the token. The use of VT-Give-Tokens forces delivery of any undelivered updates.

The VT user which does not currently possess the token may request the token from the owner by issuing a VT-Request-Tokens primitive. This has no effect on the VT service provider.

12.4 EXTENDED FACILITY SET

The extended facility set addendum to the Basic Class, which is in the early stages of drafting at the time of writing, adds to the Basic Class the ability to impose some structure to display objects.

Two types of structure are defined:

(a) Blocks, which are simple, possibly overlapping, subdivisions of the display object array. Blocks are used solely as a means for conveniently dividing up the display object and they only affect addressing operations on the display object. When blocks are used, they subdivide a page into a one-dimensional sequence of blocks. Addresses are then specified in terms of X and Y *offsets* from the start of the current block B. A display pointer then has the form (X,Y,B,Z). An example of a page subdivided into blocks is shown in Fig. 12.6.

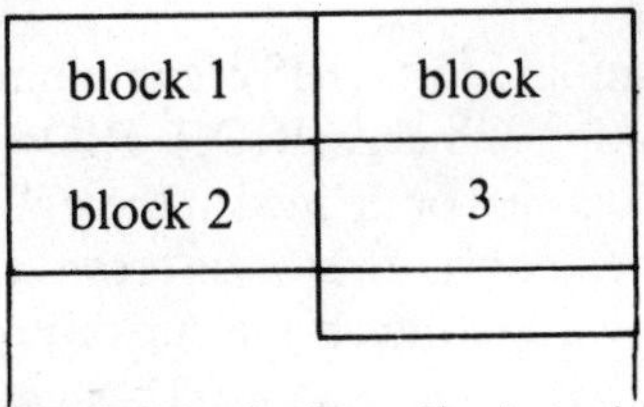

block 1	block
block 2	3

Fig. 12.6 Blocks in a display object.

(b) Fields, on the other hand, must not overlap and are intended to be used for the validation and control by the terminal of data entered by the (human) terminal user. This extends the basic class to cover a number of commonly found terminals which can locally, i.e. in the terminal, validate data as it is entered. The main advantages are:

(i) response to invalid data entry is fast

(ii) communications capacity is not wasted on incorrect data

(iii) computer resources are saved by not having to check for incorrect data from the terminal

An 'Entry Rules Control Object' (ERCO) can be defined in the conceptual communications area by one VT user which defines how the peer VT user should control the entry of data into fields. The control of data entry is performed only by the VT user and is not visible in the OSI environment so it is not a protocol violation error if the data entry is not controlled according to the definition contained in the ERCO. A field may consist of one or more (not necessarily adjacent) rectangular areas. A field must be fully contained within a page. When a display object has been negotiated to contain fields, it is termed in the addendum to be 'logically structured'. Each page in a logically structured display object is subdivided into a sequence of fields and is addressed via a field pointer, F, and an offset within the field, N. The current values of F and N define a 'logical pointer' which may be updated by addressing operations as described below. The field description in the SIS defines the mapping from (F,N) pairs into the (X,Y) pairs in the page.

There are some restrictions imposed on the Basic Class facilities when using the extended facilities. First, when blocks are used both the X and Y dimensions must be bounded. Second, windowing cannot be used in the X, Y or B dimensions.

12.4.1 Model extensions

The addendum extends the VT Model to include, in the conceptual communications area:

(a) Reference Information Store (RIS): containing none, one or more Reference Information Objects (RIOs). RIOs are display or control object updates, predefined or defined at the start of the VT connection, which are used, for example, for storing error messages or for updating the control object which controls the cursor position.

(b) Structure Information Store (SIS): containing the descriptions of each of the fields and blocks in a display object. There is one SIS for each structured display object. The display objects and corresponding SISs are held within the conceptual data store of the conceptual communications area.

(c) Entry Rules Control Object (ERCO). This is a new type of control object which may be used only when logical structuring is agreed. The ERCO contains a sequence of Field Data Entry Rules (FDERs), each of which defines the data entry rules for one field. The FDER specifies not only the rules for correct and incorrect entries into the field but also what action to take when an incorrect entry or other event has occurred. The following are events that might need a reaction:

(i) end of field reached;

(ii) 'back-space' key pressed when at the start of a field;

(iii) entry waiting time expired;

(iv) data (i.e. not 'back-space') entered after an invalid entry. A number of 'reactions' to these events are specified. Other reactions are possible but must be handled locally. The standard reactions are:

(v) transferring to the peer VT user the successfully entered fields;

(vi) notifying to the peer VT user the event which has been detected;

(vii) reading or calling an RIO record (in order to display a message and/or to move the cursor, for example);

(viii) ignoring the event;

(ix) releasing the VT connection.

(d) Context Control Object. This is a new type of control object associated with the ERCO. The CCO performs two main functions. The computer may use it to define to the terminal which field should be entered first. The terminal may use it to inform the computer of the reason for termination of data entry.

12.4.2 Extensions to the operations

The extended facility set provides a number of new operations related to blocks, fields and the control of blocks and fields.

(a) Addressing operations performed optionally as part of a display object update are extended to provide:

(i) 'next block';

(ii) 'previous block';

(iii) 'next field';

(iv) 'previous field';

(v) 'pointer absolute' by (F,N) values, i.e. move to the N-position within field F;

(vi) 'pointer relative' by (F,N) values, i.e. move to a position defined by an offset from the current field pointer;

(vii) 'logical home', i.e. move to (1,F) in the current field.

(b) Create, modify, erase and delete operations are provided for both blocks and fields. These are all provided as display object update operations.

The create and delete operations are applied to the current block or field as defined by the corresponding pointer. Note that when a field has been deleted, it is no longer possible to relate the display pointer (F,N) to the (X,Y) position within the page. Thus, after deleting a field, no display object updates are permitted except for those which create a field or move the display pointer.

When a field is created or modified, the parameters of the operation may specify secondary attributes for the field and if it does then these secondary attributes are written to the entire field.

When a field or block is erased, the assignment of primary attributes to each character block position of the field or block is cancelled. A subparameter of the erase operations specifies whether the secondary attributes are to be modified or not. Variations are provided as for the other erase operations, i.e. erase back, forward or the entire object.

When a field or block is deleted an erase operation is performed on the whole field or block and its description in the SIS is cancelled.

(c) RIOs may be updated by the operations described below. RIOs are all named and are referenced by name for each operation. Only RIOs which are not predefined can be updated.

(d) Erase RIO cancels the contents of all records of the named RIO.

(e) Erase Record cancels the contents of a named record within a named RIO.

(f) Create Record creates a new record within a named RIO and assigns to the new record a name and contents. The contents of an RIO record are a set of display or control object updates which are stored until called up by 'Read Record' or 'Call Record' operations.

The updates stored in RIO records can be applied by the following operations.

(g) Read Record performs all the updates in a named record.

(h) Call Record first saves the display and logical pointers, then performs all updates in a named record and finally restores the display and logical pointers.

(i) Update Control Object operations are made available in order to define the contents of the ERCO and the CCO.

12.4.3 Parameters

Additional parameters are provided to enable the use of blocks and fields to be negotiated and the size of the RIO and SIS to be specified.

There is one parameter which determines a choice of two modes of transmission of data entry when using logically structured displays. The first mode causes all data from all fields to be transmitted and the second mode causes only changes to be transmitted at the end of data entry.

12.5 PROTOCOL

The VT service is provided by a VT implementation by means of protocol data units (PDUs) exchanged between end-systems.

The VT protocol data units (PDUs) are conveyed as parameters of CASE or Presentation service primitives. In general, the VT PDUs correspond exactly to the VT service primitives. There are, however, some VT PDUs with no exact correspondence in the VT service and these are highlighted below where appropriate.

VT associate service elements are conveyed as case A-Associate PDUs such that:

(a) the user information parameter carries the VT associate PDU and all its parameters

(b) the Result parameter carries the Result parameter of the VT Associate (see above description of VT associate service).

When the S-mode is required, two-way-alternate services are requested from the Session service (via the Presentation service). When the A-mode is required, two-way-simultaneous services are requested from the Session service. The Session service data token is used as the WAVAR token.

VT Resease services are conveyed in P-Typed-Data (which is not subject to token control when in S-mode). The VT Release service is mechanised in the VT protocol by a pair of VT PDU exchanges. First, a VT-Release-Request is exchanged corresponding to that described in the VT service and carried by P-Typed-Data primitives. If both VT users are happy to release the VT connection, VT-Terminate-Request and VT-Terminate-Response PDUs are exchanged, carried as parameters of the case A-Release service elements.

VT-U-Abort services are mapped into A-Abort and VT-P-Abort services are mapped into A-P-Abort. The user information of the CASE primitives carry the VT PDUs and the Abort Source parameter of the A-Abort takes the value 'CASE user'.

Updates are transmitted using the normal (token controlled, where appropriate) P-Data primitives except for updates to control objects which are not delivery-controlled and these are carried by P-Typed-Data primitives (which are *not* subject to token control). The sending VT implementation can choose to group together updates and transmit at any convenient time providing that the order of updates is maintained. The exceptions to this are when a controlled object has to be updated, the object being updated has a trigger characteristic or when VT-Deliver, VT-Give-Token or VT-Release are requested by the VT user. In these cases updates not yet transmitted must be transmitted. The receiving VT user applies received updates to the relevant objects according to the delivery control in force, if any, or when VT-Deliver, VT-Give-Token or VT-Release indications are received or otherwise at the convenience of the implementation.

The VT-Give-Token and VT-Request-Token service elements are conveyed as P-Token-Give and P-Token-Please respectively with the value of 'data token' specified for the tokens parameter.

The only other services besides the update (VT-Data) service element conveyed by PDUs contained in normal P-Data primitives are the invite, offer, accept and reject negotiation service elements of the multiple interaction negotiation service. All others are conveyed by P-Typed-Data primitives.

12.5.1 Conformance

The VT protocol states 'static' conformance requirements, i.e. requirements on the facilities offered by any 'conforming' implementation. These requirements are defined in order to improve the likelihood of achieving interworking between any two implementations.

The static conformance requirements of an implementation are that it should:

(a) support the VT-A subset;

(b) support at least one of A-mode or S-mode (although clearly if it supported neither it could not communicate at all!);

(c) support the default profile(s) defined in the protocol for the supported mode(s).

(d) be capable of accepting all valid protocol sequences from the remote implementation and responding to all invalid sequences;

(e) support the basic encodings defined in ISO 8825 (ASN.1) and it may support others besides.

It is also stated that support for any character set or character set combination designatable by ISO 2022 will be considered in conformance with the character repertoire requirements. This does not preclude the use of other character sets or even make support of the stated character sets obligatory.

Claims for conformance must be accompanied by statements of which modes, subsets, parameter ranges and registered profiles are supported and a statement as to whether the implementation can initiate or accept VT-Associate protocol elements (or both). These statements accompanying claims for conformance ensure that buyers of implementations are made aware that there can be incompatibilities between two conforming implementations.

PART 5

WIDER ISSUES

Management

This chapter describes the architecture and Standards being developed for OSI management. At the time this chapter was written (June 1986) the work was very far from stable. None the less, some of the key ideas can still usefully be presented.

13.1 BACKGROUND AND HISTORY

The Basic Reference Model identifies three aspects of OSI management standardisation:

(a) Layer management; operation within a layer to manage the activity of that layer. This is specific to one layer, both in terms of what is managed, and in terms of the protocol for remote management. The work is assigned to the groups producing the main Standards for the layer.

(b) Application management; managing (remotely) the running of processes on different computer systems; this is largely a scheduling and synchronisation matter (early work on CCR was done under this banner).

(c) systems management; managing, across all layers, the resources (hardware, software, simultaneous tasks, etc.) concerned with providing OSI communications.

Note here that protocols for management activity *not* related to the provision of OSI communications, e.g. managing a flexible manufacturing plant, or managing a distributed team of software programmers, is not part of OSI management. These are 'straightforward' application-specific protocols.

The precise requirement for protocols in the OSI management area was not clear from the Reference Model.

The individual layer groups were, until recently, too busy on their 'main' protocol to worry about management facilities. The group responsible for systems and application management attempted to develop a 'management framework' to amplify the Reference Model, and to more clearly identify the work to be done.

Meanwhile, three additional pieces of work were progressed by the group.

The first was CCR (see section 9.3), which was taken out of the management group and treated as a Common Application Service Element (CASE) at a late stage in its development.

The second was the Common Directory Service (CDS) which is still being progressed by the management group, but is generally thought to be nothing to do with OSI management! We have not covered directory Standards in this book.

The third was work under a heading called 'Control of Application Process Groups' (CAPG). This was concerned with the such basic problems as: 'I want to get six programs running on different computers at the same time, so that they can interwork'. This is largely a cross-machine scheduling problem. For programs which can run in the foreground it is typically done today by having six terminals side by side and logging each one in to one of the remote machines. For programs which cannot run in the foreground, there are no current solutions. This work (the only remaining work on 'application management') is now being handled within the SASE standardisation group, but it is essentially marking time pending completion of a more general architectural study of the support needed for 'distributed processing' in OSI. It is not discussed further in this text.

The main work on the 'management framework' is discussed in the next section, and the areas of OSI management standardisation which it identifies are discussed in later sections. None of this work is yet stable.

13.2 THE OSI MANAGEMENT FRAMEWORK

The OSI management framework says a little on the 'layer management' issues.

Layer management is said to be concerned entirely with protocol *within a layer* to aid the operation *of that layer*, *excluding* normal operations in support of peer-to-peer communications.

Examples of layer management requiring a layer management protocol are:

(a) testing the functionality of the layer below, requiring a specific protocol in the layer;

(b) transmitting parameters affecting the layer, e.g. 'take no more incoming calls', information about Quality of Service (QOS) issues, and so on);

(c) conveying error information or diagnostic data related to the layer.

Layer management also includes management concerned with *a single connection* in the layer, which may be handled by protocol elements in the normal protocol for layer operation. An example of this is statistics and charging information transmitted with a disconnect.

One final 'principle' of layer management: management protocols in a layer are concerned only with operation of the layer. Thus, they provide their services to management processes associated with the operation of the layer. They do not provide them as services to be used by protocol in the next-higher layer.

The current expectation is that significant layer management facilities will be developed and become Standards over the latter part of the 1980s, probably covering the Data Link, Network, and Transport Layers.

The remaining concern of the management framework is the provision of architecture for systems management. It should be noted that any resources which are managed by layer management can, at least in theory, also be managed by systems management. Thus, the concept of the management information base (MIB), whilst developed mainly to support systems management, is a concept which may be used also by layer management.

Figure 13.1 shows information flows into and out of the management information base. We see that information can be collected from normal protocol, which can also be influenced by MIB contents. Systems

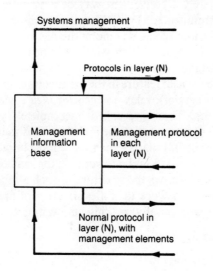

Fig. 13.1 The MIB.

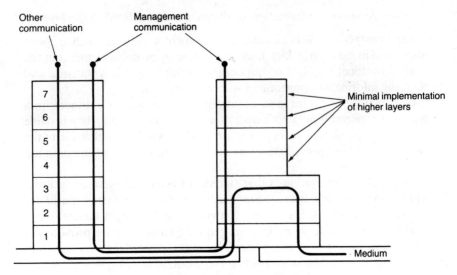

Fig. 13.2 Management of relays.

management protocol and layer (N) management protocol both read and write the MIB in a general way, except that layer (N) management protocol is restricted to those parts of the MIB concerned with layer (N).

It is important at this stage to recognise that, as with most OSI work, the *internal* handling of the MIB is not specified. It is, rather like the FTAM virtual filestore, merely a convenient way of describing the information elements to be communicated. Typical implementations of the MIB might well distribute its various elements among the pieces of code handling the layer protocols. No *database*, as such, would then be required. Alternatively it may be held as an identifiable table, file or database.

There is one other crucial feature in the architecture. This is the handling of systems (such as network relays) which *for their main function* do not have all seven layers. These are called 'incomplete' systems. A much-discussed (but now agreed) part of the architecture is that, if such systems are to take a full part in OSI management, they must implement at least the bare minimum of Layers 4, 5 and 6 to support the management protocols in Layer 7. Figure 13.2 shows the structure of such a relay.

The management framework identifies six areas of management which will be covered by (conceptually) reading and writing the MIB. These are:

(a) fault management;

(b) accounting management;

(c) configuration management;

(d) performance management;

(e) security management;

(f) name registration management.

A service and a protocol Standard are being developed for MIS – broadly, interaction with the MIB by Layer-7 protocol.

This is a multi-part Standard. Part 1 is a general overview, Part 2 is concerned with general-purpose MIB elements (and protocol support) which are of common utility, and will be used by the specific areas. This is the 'Common Management Information Service' (CMIS). Parts 3 to 8 cover MIB elements, service and protocol for the specific areas.

13.3   COMMON MANAGEMENT INFORMATION SERVICE (CMIS)

CMIS recognises four major functions, in relation to further detail of the MIB.

The CMIS work recognises that:

(a) 'OSI resources' (typically protocol handlers and links) generate 'events' which are visible to the MIB; the precise identification of 'events' (a packet transferred, an octet transferred, a CRC error, a connection established) is for more detailed future work;

(b) 'events' will be defined by the standardisation work, and will be given unambiguous names;

(c) 'events' can be 'logged' in the MIB by specifying the start time for the logging and an 'event filter' which is a list of the names of events which are to be logged; the event log is an 'attribute' of the resource, which can be read;

(d) 'events' can be 'counted' by a counter in the MIB by specifying the event type to be counted; a counter, once established, is an 'attribute' of the resource;

(e) 'aggregate counters' can be established; these include a filter specifying the 'events' they count, together with a bit-map showing whether a particular event has occurred;

(f) a counter can have a 'threshold' established, which causes a new named event to be generated when the threshold is crossed.

Within this model, service primitives provide for automatic reporting of events (which may be real ones or may be thresholds crossed), and for reading and resetting the 'attributes' of a resource, which may be counters, logs, aggregates, or parameters controlling its operations. Other 'attributes' which can be read are the 'status' of the resource, these can also be written to set the resource to a known state.

The basic service provides for:

(a) remote reporting of events;

(b) reading attributes;

(c) setting a characteristic (a characteristic is an attribute which can be set, and whose setting affects the operation of the resource);

(d) setting a characteristic if and only if the previous value was in certain specified ranges;

(e) demanding 'actions' of the resource;

(f) combinations of the above.

This then, is the CMIS service. It is highly *generic*: i.e. it is a set of potentially useful tools which will have meaning only when 'event ids', are assigned to defined events, actions, and characteristics of identified resources.

13.4 FAULT MANAGEMENT

This work has a fairly clearly defined 'requirements' annex (which will probably be omitted from the final Standard) which defines the aims. The rest of this Part is not yet sufficiently stable to report on.

Much of this work is already fully covered by the CMIS work; in particular:

(a) spontaneous error reporting;

(b) error threshold alarm reporting;

(c) continuous monitoring.

Two other main areas are being discussed. The first is confidence and diagnostic testing, and the second is a 'probe' facility to send messages round a given list of systems to check that everything is still working.

There are still some gaps between the 'requirements' annex and what is in the Standard. These are covered by terms like 'reinitialisation', 'down-line load', and counter operations such as 'wrap' or 'latch'.

13.5 ACCOUNTING MANAGEMENT

As for Part 3, this work is still in its early stages. The main function being addressed is the reporting of accounting events, with some progress on identifying when these events occur and the information associated with the event.

There is recognition of Network Level accounting events, containing:

(a) number of packets sent/received;

(b) number of octets sent/received;

(c) time of day;

(d) duration of the connection.

There is also recognition of Application Level accounting covering activities in Layers 4 to 7 of the two end-systems.

There is also discussion on providing for control of *where* accounting events, when generated, are to be reported to (the initiator of the connection, the responder, or a third party).

13.6   CONFIGURATION MANAGEMENT

Work here is also at a very early stage. The current draft document concerns activating and deactivating of resources, third-party control of connection establishment or disestablishment, downline loading, and initialisation. (Note that there is some current overlap with fault management.) We are unlikely to see a Standard in this area before the 1990s.

13.7   PERFORMANCE MANAGEMENT

This area is concerned mainly with controlling the collection of (historical) statistics (and, hence, with *summaries* of events) on a long-term basis. It is too soon to say what will actually be in the Standard.

13.8   SECURITY MANAGEMENT

The work of the management group in this area is marking time until the security addendum to the basic Reference Model is in place.

Meanwhile, the ISO Technical Committee concerned with banking is producing a Standard on 'Encryption key management', and has proposed it as a component of OSI security management.

The work is stable and very thorough. It emphasises manual procedures as well as use of communications lines. It specifies heavy use of dual control and relies on manual distribution of a master key, which is either used to distribute data keys, or to distribute submaster keys which themselves are used to distribute data keys.

The work covers both simple point-to-point operation (A and B have manually exchanged master keys) and also operation using both a key distribution centre and a key translation centre.

For both these modes, each user has only *one* manually distributed key for *all* his communications requirements. This is the key which enables him to communicate with the key distribution or the key translation centre to obtain suitable keys for communicating with other parties.

In the case of the key distribution centre, site A would request from the centre a data key for communication with site B. The centre generates a key, and passes it (encrypted) to A, together with a suitably encrypted version (using B's master or submaster key) which A can send to B when he opens the communication.

In the case of the key translation centre, party A generates a data key himself, passes it to the translation centre (encrypted with A's master or submaster key), and gets back a 'translated' version – one encrypted with B's master or submaster key, which he can then send to B.

The Standard does not quite specify a full protocol in the normal OSI sense, and does not quite fit into the OSI Presentation Layer/Application Layer architecture. It defines messages, their semantics, and the sequence in which they are used, but it does not specify how they are carried. Compared with 'normal' OSI protocols, they have a fairly rigid structure (fixed length fields, everything ASCII-encoded). These defects are, however, easily remedied, and an OSI protocol based heavily on this work seems possible.

13.9   NAME REGISTRATION AND MANAGEMENT

This is an area where there has been almost no work (as at June 1986). People are waiting to see what more needs to be done once the Common Directory Service work is complete.

13.10   CONCLUSION

The area of management in OSI has been one of continuing difficulty, largely due to lack of agreement on what needed to be done.

The year 1985/86 was, however, the turning-point when significant progress was made towards defining the form and general content of the Standards to be produced. The work will, however, continue well into the 1990s.

CHAPTER 14

Security

14.1 INTRODUCTION

Many computer systems hold data which is sensitive to some extent. University computers, for example, can hold personnel data and examination results. Military computers can hold information which is vital to national security.

When connected to communications networks these computers will be exposed to potential threats from a very wide population. The technology to communicate with, and gain access to, these computers is now available in high street shops and there is a growing band of expert 'hackers' who possess and share expert knowledge of computer systems in order to break-in to computer systems for 'fun'. Whilst these hackers are not necessarily a threat in themselves, their expertise could easily be misused by those with criminal intentions.

Thus, security in OSI will become important to almost all users of communications networks.

The question has to be asked as to whether it is sensible to standardise security mechanisms. Many people feel that by keeping their operating procedures and in particular their security precautions confidential they are improving their security since fewer people know how to attempt to breach that security. This leads to the view that security enhancements to the OSI architecture and protocols should not be standardised.

However, it is the case that the most determined attacks will come from those who will be able to find out what these operating procedures are, and thus good security will be obtained only if the security measures can withstand determined attack from a knowledgeable intruder.

Thus, I see no problem with the concept of standardising security procedures.

It is a fact that the addition of non-standard security procedures to an otherwise OSI system makes the whole system non-standard and negates the advantages of using standard OSI systems.

14.2 TYPES OF PROTECTION

The aim of the OSI security work is to ensure that a secure system can be constructed using OSI communications protocols such that the external communications are no less secure than the end-systems or any other resource.

This section will list the types of security protection or 'services' that can be required by a pair of communicating applications.

These are:

(a) Confidentiality: the protection of transmitted data from accidental or deliberate disclosure to unauthorised persons. Confidentiality is subdivided into a number of categories as follows:

  (i) confidentiality of selected fields in an SDU;

  (ii) confidentiality of a single connectionless SDU;

  (iii) confidentiality of all SDUs on a connection;

  (iv) traffic flow protection.

Traffic flow protection ensures that an observer cannot gain any useful information from observation of the number, path or frequency of data transmitted as opposed to the data itself.

(b) Integrity: the assurance that data received are exactly as sent by an authorised entity, i.e. contains no duplications, insertions, modifications or replays.

  As with confidentiality, integrity is subdivided as follows:

  (i) integrity of selected fields in an SDU;

  (ii) integrity of a single connectionless SDU;

  (iii) integrity of all SDUs on a connection.

(c) Peer entity authentication: the identification of a remote entity (this requires that there be a means to detect a simple replay of a previous authentication sequence).

  This function does not cover the machine-identification of humans, e.g. by finger-printing. It only covers identification of a remote entity that is trying to communicate via OSI protocols. When a human is using a terminal to access a remote system the identification of the human to the local VT application is a matter which is outside the scope of OSI.

(d) Access control: the ability to limit and control the access to host systems and applications via communications links. To achieve this, each entity trying to gain access must first be identified, or authenticated, so that access rights can be tailored to the individual.

  Alternatively the intending user may possess 'credentials', the presentation of which is sufficient to grant requests for access to resources.

(e) Non-repudiation: giving protection against the recipient of data later denying that it was received or the sender later denying that it was sent.

(f) Data origin authentication: the provision of assurance that the source of data received is the one claimed. This can be important, e.g. when receiving commands and instructions.

14.3 PROTECTION MECHANISMS

This section will present the protection mechanisms relevant to OSI and which are expected to be incorporated into OSI protocols. Clearly, many means can be used to protect systems, only some of which will be visible in the communications procedures between open systems. It is only the latter which are discussed.

One of the most powerful techniques is encipherment and this can be used not only to give confidentiality but can also play a part in providing a number of other protection services. This section is not restricted to consideration of encipherment, however.

14.3.1 Passwords

Passwords can clearly play a part in authentication. In order that they cannot be monitored, great care should be taken to disguise them, possibly by using encipherment. Their advantage over encipherment-only-based mechanisms is that they do not cause great system overheads in their use. Thus, a system with no confidential data can protect itself against misuse by insisting on entry of a password before access is granted. Because passwords are not entered frequently they can be software-enciphered before transmission without too great a system overhead.

Another advantage is that passwords often need changing less frequently and therefore the management support is less burdensome.

Passwords have some specific weaknesses which are exploited by 'hackers', however. These relate to the weaknesses of the humans who use them. People find it difficult to remember passwords and get round this problem in two ways.

The first is to choose a password that is easy to remember and these very frequently are things like the names or birthdates of close family.

Where the user does not choose his password and is given a meaningless string of letters or numbers, he frequently writes it down!

14.3.2 Traffic padding

Traffic padding and spurious message-generation can be used to disguise traffic flows and possibly to remove some covert channels.

14.4.3 Manipulation detection codes

Manipulation Detection Codes (MDCs) are used to detect deliberate modification, insertion, deletion or replay of data from a data stream. Again, to be really effective against a knowledgeable attacker, MDCs should be protected by encipherment, or better still cryptographically generated and, in addition, supported by a recovery protocol.

Even when protected by encipherment, special measures must be taken to detect simple replay of a previous good message. This sometimes involves incorporation of sequence numbers and time-stamps into MDCs.

14.3.4 Physical security measures

Security measures such as locks and secure rooms and cabinets will always be necessary as part of a completely secure system. Physical security is costly to provide, so the objective of security architectures is to minimise the need for it.

14.3.5 Security audits

These can be used to analyse breaches of security in order to identify the perpetrator and method of attack. This can have a strong deterrent effect.

14.3.6 Security alarms

Alarms which alert a security centre whenever suspected attacks occur, such as a detected manipulation or repeated entry of invalid passwords, can be used to catch attackers and again can have a deterrent effect.

14.3.7 Digital signatures

Signatures which are a cryptographically-derived digest of a piece of data, can be used to protect against repudiation, i.e. later denial that the data was sent, by the sender of data. Asymmetric (public key) ciphers are used to create digital signatures so that the key used to create the digital signature is held only by the sender but the key used to verify the signature can be widely distributed.

14.3.8 Notaries

These are trusted third parties who can be used to give assurance of the origin of data and its contents.

Using this technique, all communications between two entities are conducted via the notary, who records the transactions. When data is received via the notary, the recipient will accept its stated origin and contents.

The notary's record can be used later to give proof of the transactions should one or both parties get into a dispute.

14.3.9 Trusted functionality

Trusted functionality is the name given to hardware and/or software which a user has assured himself can be trusted to carry out its tasks without jeopardising system security.

All the preceding mechanisms implicitly assume that their implementation can be trusted.

For example, it is no use having an encipherment device which occasionally, due to software error, transmits data in clear.

There are many techniques that can be used to establish trust in hardware and software. These range from thorough testing to the use of formal mathematical proofs of correctness.

OSI is not concerned with the method used to establish trust. It does, however, aim to minimise the amount of trusted software needed to implement a secure system.

14.4 ARCHITECTURE – PLACEMENT OF SECURITY SERVICES

The OSI security addendum to the Reference Model (ISO 7498: Part 2) identifies:

(a) which of the security services and mechanisms are relevant to OSI standardisation;

(b) placement of the relevant services in the layers in such a way as to enable secure communications to be achieved cost-effectively using OSI, and giving guidance to the various working groups as to what addenda are needed to layer services and protocols in order to achieve this objective.

Figure 14.1 shows a matrix illustrating the allocation of security services to the OSI layers. Where a service is shown to be present in a layer, this indicates that the particular layer is directly responsible for the achievement of the service. Clearly, in these cases the higher layers must also offer the service to their users but they will achieve that protection service by passing the request on down to the lower layers.

Service	*Layer*						
	1	*2*	*3*	*4*	*5*	*6*	*7*
Peer entity authentication	N	N	Y	Y	N	N	Y
Access control	N	N	Y	Y	N	N	Y
Confidentiality	Y	Y	Y	Y	N	Y	N
Traffic flow security	Y	N	Y	N	N	N	Y
Integrity	N	N	Y	Y	N	N	Y
Data origin authentication	N	N	Y	Y	N	N	Y
Non-repudiation	N	N	N	N	N	N	Y

Key: Y – Yes, the service will be included in the layer as an optional service

N – No, the service will not be provided within the layer

Fig. 14.1 Matrix of security services and OSI layers.

14.5   NON-OSI ASPECTS

A number of important security measures are currently deemed to be outside the scope of OSI standardisation as has been indicated earlier.

This does not necessarily imply that no international standards will be produced to cover these aspects. It only means that they have no visibility in the OSI environment and therefore have no effect on the OSI protocols used between two end-systems.

One of the prime examples of this is encipherment algorithms. These *are* being standardised by ISO independently of OSI. However, the OSI architecture should be such that any suitable encipherment algorithm can be used at the defined places in the architecture and it therefore does not specify the algorithms to be used.

The following security measures are outside the scope of OSI:

(a) Physical security, e.g. locks and guards;

(b) Personnel identification methods, e.g. finger-printing or badge-readers;

(c) Operating system security, i.e. ensuring rigorous separation between software modules;

(d) Security audit analysis programs;

(e) Suppression of electro-magnetic radiations.

14.6 SYSTEM SECURITY

When planning a secure distributed system, a number of steps must be taken before the most appropriate security measures can be determined. OSI security services and protocols needed can only be determined as part of this overall security risk analysis.

The risk analysis will:

(a) Determine the 'cost' to the application of each possible type of risk. Risks include disclosure of information, modification, unauthorised access to assets, denial of service and destruction.

Assets are all the items of value to the user including the system hardware, supporting services such as air-conditioning and data.

As an example, consider a computer system providing a directory service for a distributed network. As all network users are allowed to use the directory, access and disclosures are not problems. On the other hand, denial of service and destruction could seriously impact the effective use of the whole network and are therefore of great importance.

As a further example, a banking application performing funds transfer will be most concerned about modification and destruction of data and masquerading by unauthorised users. It will not be concerned (at least to the same extent) with protection against disclosure of information.

(b) Determine the value to be derived from breaking security by a possible attacker. The values assessed will depend on the type of attacker.

For example, if the system contains commercial secrets, then the value to a rival organisation may be considerable and thus sophisticated and determined attacks may be mounted. If the system contains examination results then the attackers may be determined but less sophisticated.

This exercise will determine the level of threat, which in turn may be used to assess the likely frequency of attack.

(c) Consider the system's weak points. These could include physical access to terminals and other equipment, radiations, system software loopholes, use of broadcast technology (satellites or local area networks, etc.), subversion of staff, wire-tapping (active or passive), mutual distrust of communicating parties (where contracts are involved) etc.

For example, if the system contains commercial secrets but resides entirely in a secure building, then the only form of external threat is from radiation. If the inter-system communications are considered to be weak points, e.g. attacks via other equipment on the same network or attacks via line-tapping or equivalent, then OSI security measures must be considered.

Thus, it can be seen that OSI security has to be considered as just one part of an overall system security policy. Factors such as security classification of 'objects' and management of access rights of 'subjects' and permissible actions of subjects on objects have not been treated since they will be of concern only to the applications within end-systems and are outside the scope of this book.

There are a number of publications on the subject of risk analysis which cover the subject in much more detail. One of these is the FIPS Pub. 65 *Risk Analysis of ADP Systems*.

14.7   PROTOCOL ENHANCEMENTS

The work of ISO aimed at achieving an architecture for secure OSI communications has been discussed above. This is only a start for the real work of defining service and then protocol enhancements.

At the time of writing, the work or protocol enhancements has not made great progress. The Physical Layer Standard is at DIS stage but enhancements for the other layers have not yet got as far as establishing a base document within ISO. In the USA, ANSI have fairly advanced drafts for enhancements to the Transport and Presentation Layers, however.

14.7.1 Physical Layer encipherment

This document, currently DIS 9160, specifies the inter-operability requirements for encipherment at the Physical Layer.

The main body specifies requirements in general terms, not specific to any encipherment algorithm.

An Annex specifies the particular requirements for using the DEA-1 (DES) algorithm.

The general requirements are specified for a number of Physical Layer Standards (V.24/X.21bis, X.21, synchronous, asynchronous, etc.). For each of these, the interface conditions are defined under which an initialisation variable is sent and upon which encipherment is commenced. For full-duplex operation, two initialisation variables are sent, one for each direction of transfer; for two-way alternate, one initialisation variable is sent each time the line becomes ready for sending data.

The additional procedures for use of DEA-1 specifies a number of options for how long the initialisation variable should be. The mandatory requirement is for an initialisation variable of 48 bits. Use of 64 bits is an additional option.

CHAPTER 15

Conformance, Conformance Testing and Procurement

The objective of OSI is to enable heterogeneous systems, implemented in different and independent ways, to interwork with one another. Two issues immediately arise in trying to meet this objective. Firstly, the appropriate standard itself must be rigorous and unambiguous and only open to a single interpretation. Secondly, a prospective purchaser needs some way of determining whether a given system has, in fact, been implemented in accordance with such a standard. The definition of conformance, and the specification of conformance requirements is by no means trivial.

15.1 CONFORMANCE REQUIREMENTS

All OSI Standards contain the necessary material which determines exactly what must be implemented. Some standards contain options, for various reasons, and thus the precise conditions for use of such options must also be specified. Generally, a standard must specify:

(a) the mandatory requirements which must be observed under all circumstances;

(b) the conditional requirements which must be observed under the conditions specified in the standard;

(c) the optional requirements;

(d) the prohibitions, i.e. the standard should also specify what must not be done.

Two basic categories of conformance to standards have been identified: dynamic conformance and static conformance.

15.1.1 Dynamic conformance

A properly specified standard will specify all allowed behaviours. The set of all allowed behaviours defines the maximum capability of a system (conforming to the standard).

Dynamic conformance requirements of a standard are all those requirements (including options) which determine the possible behaviour (or set of behaviours) permitted by that standard. A particular system can be said to conform dynamically to a standard if its behaviour is a member of the set of all behaviours permitted by the Standard.

The dynamic conformance requirements are normally indicated in ISO Standards by use of the verb 'shall', e.g. 'on receipt of an XYZ PDU the receiver shall transmit an ABC PDU'.

15.1.2 Static conformance

In order to maximise interworking, certain constraints may be specified, by classifying certain requirements as mandatory for all implementations, conditional, or optional. These constraints are said to be the static conformance requirements.

The essential point about static conformance is that it is a statement of what conforming implementations should be capable of doing. Static conformance requirements are the means whereby the number of non-interworking variations of implementations of a complex standard is minimised.

From the point of view of both procurement and conformance testing it is essential that the supplier of equipment be asked to provide statements about the static and dynamic capabilities of the equipment. Such a statement is often called a Protocol Implementation Conformance Statement (PICS). Prospective purchasers of implementations should be aware that a simple claim by a supplier of the form 'the XYZ product conforms to ISO ...' will *not* comply with the conformance requirements of any OSI protocol!

A conforming system is one which satisfies both the dynamic and static conformance requirements.

It is clear that the conformance requirements need to be precise and unambiguous and, hence, there is a drive to increase the use of formal description techniques within the various standards. Formal description techniques are not discussed in this book because neither the techniques nor the method of applying them are yet stable.

In order to be sure that two or more systems will interwork a comparison of respective conformance statements (PICS) is necessary. Suppliers should not only be asked to provide statements about static and dynamic capabilities, but additionally as much complementary information as possible. This complementary information should include details regarding:

(a) non-implementation of capabilities if the purpose of the system justifies such limitations, e.g. in a system which only ever needs to receive user data and never needs to send it, it may be possible to omit the capability of sending it;

(b) additional mechanisms claimed to fix known ambiguities or deficiencies not yet fixed in the standards or in peer real systems, e.g. solutions of multi-layer problems;

(c) selection of free options which in the standards are not taken into account in the static conformance requirements;

(d) retained values of parameters when the standard allows a range, etc.

Some of the above elements may not fall within the scope of conformance testing *per se*, in which case a value judgement is necessary. Some elements, for example, may affect performance rather than conformance. The relative magnitudes of timers may also affect performance and, in some cases, prevent interworking.

15.2 CONFORMANCE TESTING

One way of determining the likelihood of two systems interworking with one another, is to subject both systems to 'live' tests.

There is no intention, at least for the present, that suppliers must, mandatorily, submit their equipment to a conformance testing authority for validation. National regulations or procurers of equipment are, of course, free to insist that this indeed be done, if suitable tests and testing authorities are available. Some manufacturers may also feel that some 'seal of approval' for their equipment may provide a 'marketing advantage' and thus themselves voluntarily submit their equipment for validation by a conformance-testing agency.

The voluntary nature of conformance testing, however, in no way obviates the need for standardisation for conformance testing. In order to compare properly two systems they must be subjected to the same tests. Furthermore, repeatability of tests is necessary during the de-bugging of systems, and acceptance of test results on a worldwide basis is required from different testing agencies if equipment is not to be habitually subjected to the same or different conformance tests.

It follows from this that standardisation is required in two areas: firstly, for test methods and, secondly, for the test suites for the particular methods.

ISO is progressing a document which is expected to become a further addendum to ISO 7498 (Basic Reference Model) and, in addition, groups working on individual layer protocols are producing appropriate test suites.

Many countries including the UK and the USA have test agencies in place with capability to test some of the OSI protocols. The National Computing Centre (UK) and National Bureau of Standards (USA) are two such agencies.

The OSI protocol standards define the allowed behaviour of a protocol entity with respect to the service primitives above and below the entity. Three methods of conformance testing have been identified and provide flexibility according to:

(a) whether it is the Abstract Service Primitives (ASPs) or the protocol data units (PDUs) that are to be observed and controlled;

(b) the layer to which the ASPs or PDUs apply;

(c) whether the ASPs or PDUs are controlled and observed within the System Under Test (SUT), or in a remote system under test.

Service primitives are expressed in abstract terms so as to not restrict the freedom of testers/implementors to implement the tests in different ways. Thus, use can be made of whatever interfaces are accessible, provided that the required degree of control and observation is achievable.

15.2.1 Local test method

This method involves testing the (N)-entity in terms of the upper and lower service primitives, and does not involve real communication with another system. Procurers of systems should not rely on such tests as they do not give sufficient confidence in the system behaviour in live communication. Figure 15.1(a) illustrates the principle.

15.2.2 Distributed test method

This method is illustrated in Fig. 15.1(b). This method provides a comprehensive means of testing, since the (N)-entity under test can be controlled and observed from above, i.e. via (N)-ASPs, and can be judged in relation not only by observing the (N)-PDUs but also in terms of its use of the (N-1)-layer service.

15.2.3 Remote test method

This test method places minimum requirements on the system under test. This method has been used frequently in the past for testing systems, and the tests are performed solely by means of the PDUs being exchanged between the SUT and the test system (see Fig. 15.1(c)).

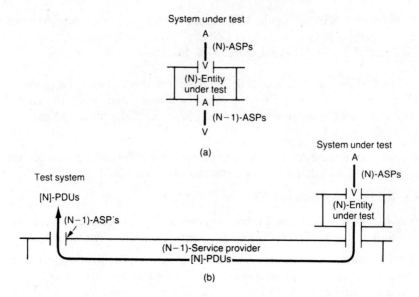

(a)

(b)

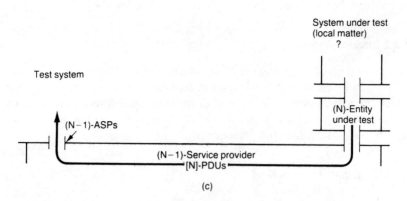

(c)

15.1 Test methods; (a) local test method; (b) distributed test method; (c) remote test method.

Multi-layer testing is possible for all three methods, and does not require access to all SAPs, i.e. does not require access to SAPs between the layers under test.

The testing of Network Layer relays creates a number of difficulties because of the need to deal with both sides of a relay simultaneously. The lack of a single service definition below the Network Layer means that testing can only be performed in terms of PDUs and not ASPs.

Methods of describing and formulating tests for use within the three test methods are still under study.

15.3 PROCUREMENT

Several procurement issues were highlighted in the previous two sections, primarily dealing with the type of information a prospective purchaser should obtain (indeed demand) from the potential supplier. This section is concerned with the more general strategic aspects of procurement.

The development of OSI Standards within the framework of the OSI Reference Model is undoubtedly the most significant standardisation effort ever undertaken. The scope of the activity is enormous and the development is almost certainly a continuous process, especially in the application area. It is also true that different parts of the Model are progressing at different rates, producing a jig-saw effect.

If all the required OSI Standards were established and available from all manufacturers and suppliers of equipment, it would be a procurer's paradise. Systems could be procured competitively and be guaranteed to interwork with one another. The constraints on to whom any given user could communicate would be negligible. Longevity and evolution capability of systems would be assurred, and costly interworking conversions would be a thing of the past.

However, OSI has not quite reached this state of paradise yet. To a large extent, it can be argued, it is actually the procurer's responsibility to see that it does, by insisting on, and only accepting, systems with either OSI in place, or with the OSI potential. Certainly, users should not underestimate the effect of their purchasing power on suppliers' attitudes and availability of OSI products and services. Notwithstanding this, there is bound to be an interim period during which users face some difficult procurement decisions.

Two procurement strategies can be identified:

(a) the 'big bang' strategy;

(b) the incremental strategy.

15.3.1 The 'Big Bang' strategy

This strategy is based on an assumption that the necessary standards and products will take several more years to become available and, thus, it is better to wait until that time, and then make the required massive change. This strategy presupposes that partial implementation of OSI is not worthwhile, and that suppliers will also, without interim pressures, suddenly produce a complete set of OSI products and services.

There is one grave danger associated with this approach. During the interim period the user will be 'saddled' with a particular proprietary approach. This is not bad in itself, but immediately locks-in the user to the particular supplier for what may be an indefinite period, and reduces the incentive for the supplier actually to migrate towards the user's objective, and the OSI objective.

15.3.2 Incremental strategy

This strategy involves making maximum use of such OSI Standards as are completed, and adding the others as and when they become available. Thus, there will always be a gradual convergence towards the ultimate object, albeit in small steps during the interim period.

This approach will undoubtedly be painful and relatively expensive in the short term. However, the long-term benefit is the achievement of OSI, and the duration of the interim period may in fact be substantially reduced as a result of customer pressures exerted on manufacturers by way of this strategy.

The amount of effort and expertise required from the user is much greater, but so are the rewards. The well-informed user–supplier relationship can only be beneficial for the user, and his ability to procure competitively.

Careful planning, even with partial OSI implementations, can substantially increase the 'openness' of users systems, and ensure that systems evolve towards the ultimate objective. A continual pressure can be exerted on suppliers, who are forced from the outset to supply the necessary framework for the evolution.

INDEX

INDEX

A-Abort, 234
A-Associate, 231, 232
'abandon', 195
'abstract syntax', 207, 211, 214, 292
   named, 215
Abstract Syntax Notation One, *see*
   ASN.1
'access context', 269
access control, 272, 364
access control store (ACS), 331
accounting management, 360
ACS, *see* access control store
ACSE, *see* association control service
   elements
'Activities', 182, 183, 186
activity management, 196
address, 54
address parameters, 135
AFI, 127
A-mode, 331
A-P-Abort, 231, 234
'Application', 206
'Application entity', 206
Application Layer, 24, 206
Application management, 355
'Application process', 206
'Application service elements' (ASEs),
   207, 230
Application titles, 54, 55
applications, 21
architecture, 8
A-Release, 231, 233
ASEs, *see* 'Application service elements'
ASN.1 (Abstract Syntax Notation One),
   214, 221
assignment, 165
assignment rights, 177
association control, 230
association control service elements
   (ACSE), 231
asynchronous transmission, 152

atomic action, 242
'atomic action identifier', 241
A-U-Abort, 231
authentication, 320
   peer entity, 364
   third-party, 320
authorisation, 320

Bachus-Naur Form (BNF), 214
'back-pressure', 75
'Basic Class', 305
Basic Encoding Rules, 227
binary concrete syntax, 129
bit string, 222
blocking, 35
Boolean, 222

CASE, *see* 'Common Application
   Service Elements'
call redirection, 184
calling/called address extension, 130
C-Begin, 243
CCA, *see* 'conceptual communications
   area'
CCITT, 16
CCITT, Q.921 (LAPD), 133
C-Commit, 243
CCR, *see* commitment, concurrency
   and recovery
CDS, *see* Common Directory Service
checkpointing, 247
checksum, 170
CLNS, *see* connectionless-mode
   network service
Class 0, 172
Class 1, 172
Class 2, 172
Class 3, 172
Class 4, 174
clear confirmation, 91
clear indication, 91

clear packets, 95
clear request, 91
CMIS (Common Management
    Information Service), 359
COI, *see* connection-oriented
    interconnection
'collision', 178, 200
collision resolution, 178
commitment, 243
commitment, concurrency and recovery
    (CCR), 230, 236, 250, 279, 291, 303,
    321
commitment level, 324
'Common Application Service Elements'
    (CASE), 207, 230
common application services, 15
Common Directory Services (CDS), 356
Common Management Information
    Service, *see* CMIS
communication, 7
complete source routing, 109, 113
concatenation, 36, 168, 201, 202, 204,
    268
'conceptual communications area'
    (CCA), 330
concrete syntaxes, 129
concurrent access, 240
confidentiality, 364
configuration management, 361
Confirm, 31
confirmed-service-element, 32
conformance, 174, 372
conformance testing, 372, 374
connection, 7
connection end point identifier, 28
connection establishment, 41, 134, 167,
    184, 203
connection establishment delay, 73
Connection-mode Network service
    (CONS), 69, 82
connection-mode service, 157
connection-oriented data transfer, 136
connection-oriented interconnection
    (COI), 118
connection-oriented mode, 27
connection refusal, 168
connection release, 43, 161
Connection Release service, 135
connectionless data transfer, 136
connectionless mode, 27, 47
connectionless-mode network service

(CLNS), 105, 106
connectionless Transport service, 164,
    175
CONS, *see* Connection-mode Network
    service
'constraint sets', 262, 265, 275, 284, 285
'context', 207
Context Control Object (CCO), 347
context management, 279
'control objects', 330
C-Prepare, 243
C-Ready, 243
'credit', 170
'credit' mechanism, 44
credit reduction, 170
C-Restart, 243
C-Refuse, 243
C-Rollback, 243

(D)-bit, 89
data-circuit-terminating equipment
    (DCE), 86
Data Link Layer, 23, 24, 82
Data Link level, 86
Data Link protocol, 133
Data Link Service, 82, 132
data origin authentication, 365
data packet, 89
data structure definition (DSD), 331
data terminal equipment (DTE), 86
data TPDU numbering, 168
data, transfer of, 44
data transfer, 136, 150, 160, 186, 187,
    204
database, 15
DCE, *see* data-circuit-terminating
    equipment
DCS, *see* Defined Context Set
DEA-1 (DES), 370
deblocking, 35
DEC VT100, 332
decimal concrete syntax, 129
DECNET, 25
Defined Context Set (DCS), 215
'delivery control', 332
demultiplexing, 169
derived PDU, 110
'device object', 330
diagnostic field, 95
dialogue control, 183, 186
'dialogue unit', 194

digital signatures, 366
directory, 38, 54, 55, 56
DIS 1960, 370
'display object', 330
'distributed applications', 206
distributed end-system, 120
DL-Data Indication, 138
DL-Data Request, 138
docket, 289
document movement, 311
'document type', 262, 265, 275, 284, 286
domain, 123, 125
Draft International Standard (DIS), 15
Draft Proposal (DP), 15
DSD, *see* data structure definition
DSP, 127
DTE, *see* data terminal equipment
duplex (two-way simultaneous), 186
dynamic conformance, 372

E.163, 122, 128, 129
E.164, 128, 129
embedding, 42
encoding rules, 214
Encryption key management, 361
end systems, 24
'entry rules control object' (ERCO), 346, 347
ERCO, *see* entry rules control object
Error Report, 113
error reporting, 114
establishment failure probability, 73
European Computer Manufacturers Association (ECMA), 17
exception reporting, 196
'execution agency', 307
Expedited data, 73, 76, 90, 157, 160, 185
Expedited data transfer, 45, 136, 169
explicit flow control, 169
extended control parameter, 185

F.69, 122, 128
FADU, *see* file access data unit
FADU identifier, 267
FADU name, 267
fault management, 360
F-CHANGE-ATTRIBUTE, 281
F-CHECK, 289
F-CREATE, 280
F-DATA, 284
F-DATA-END, 284

F-ERASE, 283
file access, 259
file access data unit (FADU), 267
file attributes, 262
file management, 260
file transfer access and management (FTAM), 238, 259
filestore directory, 261
F-INITIALIZE, 278
F-LOCATE, 283
flow control, 44, 75, 157, 160, 203
    explicit, 169
F-OPEN, 282
F-READ-ATTRIBUTE, 281
F-RECOVER, 289
F-RESTART, 290
frozen references, 171
F-SELECT, 280
FTAM, *see* file transfer access and management
full duplex, 146, 149
functional elements, 165
'functional units', 183

gateway, 59, 65
general string, 226
generalised time, 226
generic addressing, 124, 184
generic name, 55
global addressing, 121
global titles, 38
graphic string, 226
graphics, 15

half-duplex, 146, 149, 151
half-duplex (two-way alternate), 186
half-gateway, 117
HDLC, LAPB, 137
heuristic commitment, 245
heuristic rollback, 245
hierarchical address scheme, 37
'hop-by-hop enhancement', 65, 117

I.430, 152
IA5 string, 226
IBM 3270, 332
identifier, 54
IDI, 127
implicit termination, 168
inactivity control, 171
Indication, 31

initial PDU, 110
'initiation agency', 307, 308
integer, 222
Integrated Services Digital Network, *see*
    ISDN
integrity, 364
interface data unit, 34
interface unit, 33
intermediate systems, 24, 64
International Standard, *see* IS
International Standards Organisation
    (ISO), 7, 12, 14
International Telephone and Telegraph
    Consultative Committee, *see*
    CCITT
'internet', 65, 68, 117
IS, 15
ISDN (Integrated Services Digital
    Network), 152, 155
ISO, *see* International Standards
    Organisation
ISO 646, 295, 330
ISO 2022, 330
ISO 7498 DAD2, 184
ISO 7776, 86
ISO 7776 (LAPB), 133
ISO 8022, 233
ISO 8072, 156, 160, 162
ISO 8073, 156, 165, 168, 170, 174, 176
ISO 8208, 86, 96, 97, 100, 102
ISO 8326, 183, 185
ISO 8327, 200, 205
ISO 8348, 69, 154
ISO 8348:AD1, 105
ISO 8348:AD2, 121
ISO 8473, 110
ISO 8571, 259
ISO 8602, 175
ISO 8649, 230, 231, 236
ISO 8650, 230, 236
ISO 8650: Part 2, 235
ISO 8802 (LLC), 133
ISO 8802: Part 2, 97, 99
ISO 8824, 214
ISO 8825, 214
ISO 8831, 302, 305
ISO 8832, 302, 306
ISO 8878, 85, 87, 93, 97, 100, 102
ISO 8881, 97
ISO 8886, 133
ISO 9040, 328, 336

ISO 9041, 328
ISO 6523–ICD, 129
ISO DCC, 128

J-BEGIN, 321
J-COMMIT, 321
J-DISPOSE, 321
J-END-SIGNAL, 323
J-ENQUIRE, 322
J-GIVE, 322
J-HOLD, 323
J-INITIATE, 321
J-KILL, 323
J-MESSAGE, 323
job control language (JCL), 303
J-READY, 321
J-REFUSE, 321
J-RELEASE, 323
J-RESTART, 321
J-ROLLBACK, 321
J-SPAWN, 322
J-STATUS, 323
J-STOP, 324
JTM, 237, 238, 302
JTM relays, 308

LAN, *see* local area network
LAN/WAN interconnection, 117
layer, 20, 26
layer management, 355, 357
layer service definition, 27
leased line, 95
lifetime, 114
LLC1, 97
LLC2, 97
local area network (LAN), 97
logic link control, 133
logical addressing, 124
logical channel number, 92
logical link control, 140
'logically structured', 346
'long-arc', 266

(M)-bit, 89, 103
MAA PDU, 191
MAC, *see* media access control
'macro notation', 225
'Major Synch points', 182, 183, 186, 188,
    189, 194
major synchronisation, 191
'management framework', 355, 356

management information base (MIB), 357
manipulation detection code (MDC), 366
MAP PDU, 191
MDC, *see* manipulation detection code
master, 240
maximum acceptable cost, 73
maximum NSDU lifetime, 108
media access control (MAC), 99
medium, 22
medium access technique, 133
message code, 104
MIB, *see* management information base
minor synch points, 186, 188, 189
mobile addressing, 124
modem, 146
'module', 224
monitor point, 314
more segments flag, 111
multi-drop, 134
multi-endpoint, 134
multi-endpoint connections, 41
multilink procedure, 134
multiple active contexts, 217
multiple interaction negotiation, 341
multiplexing, 40, 152, 169
multipoint, 140

(N)-Connect confirmation, 69, 84
(N)-Connect indication, 69, 84
(N)-Connect request, 69, 84
(N)-Connect response, 69, 84
(N)-Connection, 27
(N)-connection endpoint identifier, 39
(N)-Data acknowledge, 76
(N)-Data-acknowledge indication, 84
(N)-Data-acknowledge request, 84
(N)-Data indication, 75, 84
(N)-Data request, 75, 84, 89
(N)-Disconnect confirmation, 69
(N)-Disconnect indication, 74, 84
(N)-Disconnect request, 69, 74, 84
(N)-Expedited-data indication, 84
(N)-Expedited-data request, 84
(N)-Facility, 51, 107
(N)-Layer, 26
(N)-relay, 48
(N)-Report, 107
(N)-Reset confirm, 84
(N)-Reset indication, 84

(N)-Reset request, 84
(N)-Reset response, 84
(N)-SAP, 28
(N)-Service, 27
(N)-Service User, 27
(N)-Unitdata, 33, 50, 105
name registration and management, 362
named abstract syntax, 215
National Bureau of Standards, 375
National Computing Centre, 375
NCMS, *see* Network connection management sub-protocol
negotiation, 134, 167, 204
'net-effecting', 333
network administration authority, 123
Network connection management sub-protocol (NCMS), 166, 176
Network expedited, 167
Network Layer, 22
Network service, 22, 59, 61, 68
Network service data units (NSDUs), 69
node, 266
non-repudiation, 365
notaries, 367
NSAP address, 54, 56
NSDUs, *see* Network service data units
null, 222
numeric storing, 226

octet string, 222
Open Systems Interconnection (OSI), 3
optimised dialogue transfer parameter, 185
OSI, *see* Open Systems Interconnection
OSI architecture, 15
OSI management, 15, 355
OSI Reference Model, 20

P(R), 89, 101
Packet Assembly/Disassembly (PAD), 93
Packet Level, 86
Packet Level protocol (PLP), 86, 97
PAD, *see* Pocket Assembly/Disassembly
P-Alter-Context, 219
partial source routing, 109, 113
passwords, 365
PDU checksum, 116
peer entity authentication, 364
peer-to-peer protocols, 6, 7, 20
performance management, 361

Physical Layer, 23, 145
Physical Layer activation, 146
Physical Layer deactivation, 150
physical medium, 145
physical security measures, 366
Physical service, 145
PICS, *see* Protocol Implementation
    Conformance Statement
PLP, *see* Packet Level protocol
'pre-order traversal sequence', 267
'PREPARE' PDU, 191
PREPARE-RA, 192
'PREPARE-RS', 192
PREPARE SPDU, 203
Presentation context, 215
Presentation data values, 292
Presentation layer, 15, 24, 206, 210
printable string, 226
priority, 72, 134
'profiles', 336
proforma, 308
protection, 72
protocol class, 167, 171
protocol control information (PCI), 33
protocol data unit (PDU), 33, 34
protocol identification, 176
protocol indentifier, 93
Protocol Implementation Conformance
    Statement (PICS), 236, 373
protocol specifications, 27
protocol unit, 33
PSTN (Public Switched Telephone
    Network), 155
Public Switched Telephone Network, *see*
    PSTN
'purge', 203

Q-bit, 94, 103
QOS maintenance, 114
quality of service (QOS), 22, 42, 51, 72,
    134, 155, 158, 162, 184
queue model, 78

real networks, 22
reassembly, 35
reassignment after failure, 169
receipt confirmation, 73, 76, 172
recombining, 171
recording of route, 113
Reference Information Store (RIS), 346
'regimes', 276

REJ, 102
'relay units', 65
relaying, 48
relaying and routing, 64
relays, 64
release, 168
release delay, 73
release failure probability, 73
'reliable' protocols, 251
'reliable service', 284, 288
remote job entry, 302
report movement, 312
Request, 31
resequencing, 171
reset, 47, 77, 137
Reset request, 77, 90
Reset service, 69
residual error rate, 73
resilience, 73
Response, 31
'restart', 195
resynchronisation, 169, 182, 186, 192,
    195, 204
retention until acknowledgement, 169
retransmission on timeout, 171
RIS, *see* Reference Information Store
risk analysis, 369
roles, 62
'rollback', 240, 241, 243
root, 266
routing, 38
RR, 89
RS PDU, 192

S.70, 172
S-Activity-Discard, 197, 198
S-Activity-End, 197, 199
S-Activity-Interrupt, 197
S-Activity-Resume, 197, 198
S-Activity-Start, 197
SAP, *see* service access point
SASE, *see* 'Specific Application Service
    Elements'
SC6, 14
SC18, 14
SC20, 14
SC21, 14
S-Capability-Data, 187, 188, 197
S-Connect, 184
S-Control-Give, 187
S-Data, 187

security, 113
security alarms, 366
security audits, 366
security management, 361
segment permitted flag, 111
segmentation, 35, 110, 167, 201, 202, 204
segmentation/reassembly, 75
selector, 54, 55
'semantics', 207, 210
separation, 36, 168
sequencing, 47
service, 8
service access point (SAP), 28, 36
service data unit (SDU), 33
service definition, 30
service levelling, 133
service primitives, 30, 31
service unit, 33
Session connection release, 199
Session dialogue, 189
Session expedited, 193
Session Layer, 15, 24, 181
Session protocol, 200
'set', 195
S-Expedited data, 187, 188
SG VII, 16
SG VIII, 16
simplex, 146, 149, 151
'sink agency', 307
SIS, *see* structure information store
slaves, 240
S-Mode, 331
SNA, *see* Systems Network Architecture
SNAcP, *see* subnetwork access protocol
SNDCP, *see* subnetwork dependent
    convergence protocol
SNICP, *see* subnetwork independent
    convergence protocol
SNPA, *see* subnetwork points of
    attachment
'source agency', 307
'source reference', 98
source routing, 109
'Specific Application Service Elements'
    (SASE), 207
specific application services, 15
S-P-Exception-Report, 196
splitting, 40, 162, 171
S-Release, 199
S-Synch-Major, 194
S-Synch-Minor, 193

state transition diagram, 78
static conformance, 373
S-Token-Give, 187
S-Token-Please, 187
structure information store (SIS), 346
Study Groups (SGs), 16
S-Typed-Data, 187, 188
S-P-Abort, 200
S-U-Abort, 199
subcommittees, 14
subjob, 308
sublayers, 25
subnetwork access protocol (SNAcP),
    62, 63
subnetwork dependent convergence
    protocol (SNDCP), 62, 63, 85
subnetwork independent convergence
    protocol (SNICP), 62, 63
subnetwork interconnection, 117
subnetwork points of attachment
    (SNPA), 100, 122
subordinate, 240, 242
S-U-Exception-Report, 196
superior, 240, 242
'symmetric synchronisation', 193
synchronisation, 24, 151
synchronisation point serial numbers,
    185
synchronisation services, 188
synchronous transmission, 152
synonym, 38, 56
systems management, 355
Systems Network Architecture (SNA), 6,
    25

T.70, 103, 157
'tag', 224
TC97, 14
T-Connect, 158
Technical Committee (TC), 14
T-Disconnect, 159, 161
telex string, 226
third-party authentication, 320
throughput, 72
throughput QOS, 88
tilde, 32
time sequences, 31
timer T25, 100
title, 54
TLV, 228
token management, 187

'tokens', 182
TPDU ('transport-protocol-data unit'),
     165
TPDU transfer, 167
traffic padding, 366
transfer control record, 316
transfer failure probability, 73
transfer of data, 44
'transfer syntax', 208, 210, 292
transit delay, 72
transit delay QOS, 86
Transport entity, 154
Transport Layer, 23, 24, 154
Transport protocol, 164
Transport protocol class, 165
'transport-protocol-data-units', *see*
     TPDUs
Transport service, 154, 156
transport service access point (TSAP),
     158
tree of activity, 242
tree structure, 266
'trigger', 331
trusted functionality, 367
TSAP, *see* transport service access point
T-Unitdata, 164
'two-phase commitment', 236
two-way alternate working, 183
two-way simultaneous data flow, 183
Type, Length, Value, *see* TLV
Type-reference, 224

UN, 175
UN TPDU, 176
Unitdata (UD), 27, 175
'used of expedited', 135
'user-correctable service', 284
UTC time, 226

V.22, 146
V.24, 146
V.28, 146
videotext string, 226
virtual filestore, 260, 262, 271
Virtual Terminal, 327
'Visible String', 223, 226
VT-Ack-Receipt, 344

VT-Connect, 337
VT-Deliver, 333, 344
VT-End-Neg, 341
VT-Give-Tokens, 345
VT-Neg-Accept, 341
VT-Neg-Invite, 341
VT-Neg-Offer, 341
VT-Neg-Reject, 341
VT-P-Abort, 340
VT-Release, 339
VT-Request-Tokens, 345
VT-Start-Neg, 341
VT-Switch-Profile, 340
VT-U-Abort, 339

WACA, *see* write access connection
     acceptor
WACI, *see* write access connection
     initiator
WAVAR, *see* write access variable
'window', 343
'window' mechanism, 44
work manipulation, 315
work specification, 307, 308
working groups, 14
write access connection acceptor
     (WACA), 332
write access connection initiator
     (WACI), 332
write access variable (WAVAR), 332

X.3, 327
X.21, 102
X.25, 85, 170
X.25 facilities, 87
X.25 PLP, 100
X.28, 327
X.29, 93, 177, 327
X.121, 122, 128
X.213, 69
X.224, 174
X.225, 200
X.244, 94
X.400, 231
X.409, 213, 214, 221, 264
X.410, 254
Xerox 'Courier', 264